IMAGES
of America

KEYPORT FIREFIGHTING

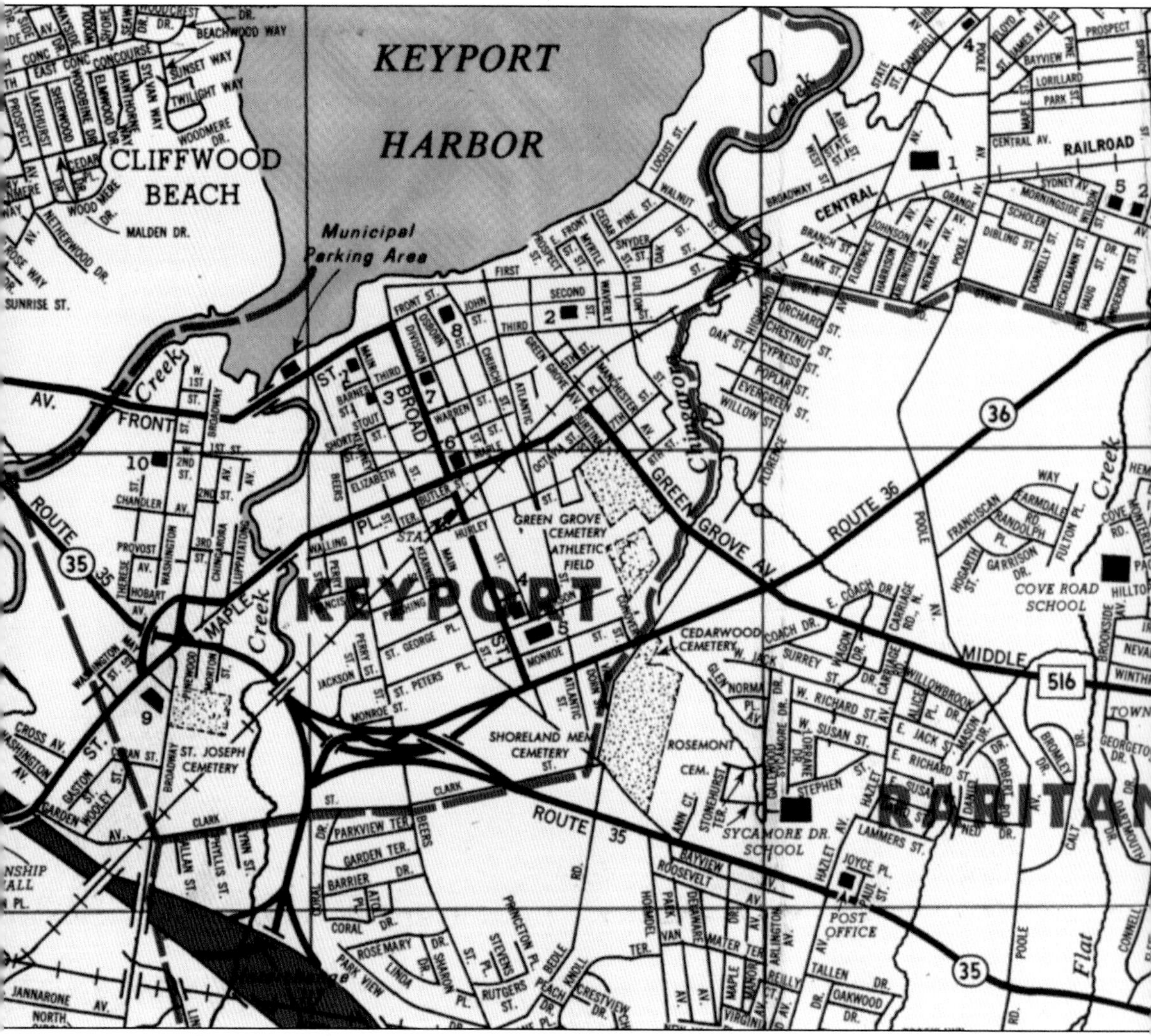

MAP OF KEYPORT. This 1960s map of the Keyport area clearly shows the compact nature of this 1.4-square-mile, nearly 200-year-old town that lies along the shores of the Raritan Bay. Over 7,000 persons call Keyport home, where the Keyport Fire Department has provided fire protection as volunteers since October 10, 1877.

On the cover: **KEYPORT HOOK AND LADDER COMPANY NO. 1.** In this 1958 photograph, the members of Keyport's oldest fire company pose with their new aerial ladder truck. On the ladder, from top to bottom, are Dorman Walling, Bob Rose, Willy Siegel, and Ed Morrissey. On the truck, from left to right, are Charles Grey, ex-chief Elgar Walling, Harry Aumack, Jules Rosato, Tony Infante, and Bob Rosato. (Courtesy Keyport Hook and Ladder Company.)

IMAGES
of America

KEYPORT FIREFIGHTING

Tom Gallo, Tim Regan, Jim McTernan,
and the Keyport Fire Museum

ARCADIA
PUBLISHING

ISBN 978-0-7385-6361-9

Published by Arcadia Publishing
Charleston, South Carolina

Printed in the United States of America

Library of Congress Control Number: 2008936755

For all general information contact Arcadia Publishing at:
Telephone 843-853-2070
Fax 843-853-0044
E-mail sales@arcadiapublishing.com
For customer service and orders:
Toll-Free 1-888-313-2665

Visit us on the Internet at www.arcadiapublishing.com

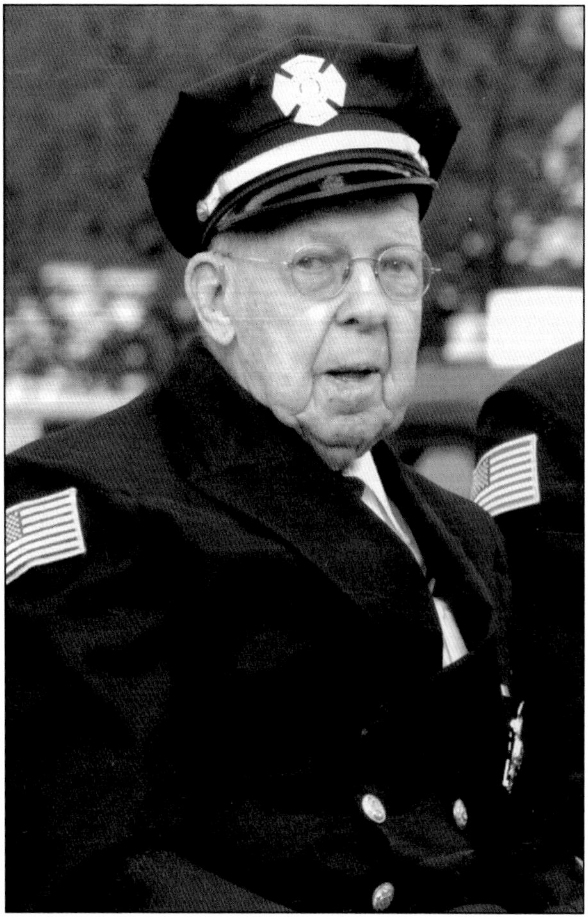

MELDON "BABE" STULTZ. This book is dedicated to Meldon "Babe" Stultz, president of Eagle Hose Company No. 4. Since joining the department in 1943 he has answered nearly 10,000 fire calls. Babe is the longest-serving active member of the Keyport Fire Department since inception. With his wife Elizabeth's support, he has achieved this remarkable record. Thank you, Babe.

Contents

Acknowledgments

The authors gratefully appreciate the assistance of so many people. The photographs now belonging to the Keyport Fire Museum and Education Center (KFMA) were donated by relatives and friends of former and current members. This single largest repository was the most important source of images. Images that have no credit lines are courtesy of the KFMA. Special thanks to Bertram Morris and Tim Regan, who donated their collections to the KFMA.

Equally as important, other contributors provided rare looks at equipment that was not often photographed. This allowed us to assemble for the first time in print the most complete grouping of all apparatus ever owned by the Keyport Fire Department (KFD). They also put names to unrecorded faces on older images. Thanks are extended to major contributors Anthony Agnello, Harry M. Aumack II, Roger Benedict, Jacqueline Brandauer, Ken Goble, Robert Holtz, William Larkin, Peter Masses Photography, Robert Poling, Gene Regan, Joel Rosenbaum, Keyport Historical Society (KHS), and Anthony Vecchio.

The authors' proceeds from this publication benefit the KFMA to further preserve Keyport's rich firefighting history. Your support is appreciated.

INTRODUCTION

Ben Franklin coined the phrase "Necessity is the mother of invention." That is difficult to debate and especially appropriate as it relates to the disastrous great fire that resulted in the formation of the Keyport Fire Department. Although there was no loss of life, the devastation of destroyed buildings and private property prompted to need for a fire department. Residents and the business community recognized the need and acted quickly to organize firefighting manpower and equipment beyond a simple bucket brigade. Meanwhile, using less flammable material, such as brick, stone, and cement, the business area of town was rebuilt. While this evolution of the fire department is true of most towns' and cities' fire companies, Keyport's story is special to its residents. This is due in part to the second, third, and fourth family generations, as well as newcomers, who continue to join, voluntarily, this critical community service.

Keyport was once a privately owned plantation sold at public auction into privately acquired lots in 1830. From that point, it grew to an important, bustling center of commerce. Jobs abounded, and its location on the bay made it an attraction for good living and raising families. The natural harbor served as a perfect point to ship goods to New York markets on accommodating piers. The tidal water allowed for heavy shipping, recreational ice-skating and swimming, pleasure craft and fishing, and oyster businesses. Keyport streets were filled with merchants and livestock, overlooked by hotel porches full of mariners and travelers alike.

This prosperity and harmony came to a standstill on the evening of September 21, 1877. A group of National Guardsmen were on their way home from a meeting when they were surprised by a sheet of skyward-bound flames behind a market store about midway up the hill on Broad Street. The guardsmen knew saving this building was futile, acting next to alert as many people as possible of the danger in hopes of savings lives inside adjoining buildings. Soon docked steamboats sounded their whistles while church bells rang to alert all to wake up and save themselves—and to assist! Attempts were obviously having little to no effect as the fire spread to each contiguous building naturally made of wood in this time period. Fed by winds, the fire was intense and wild, jumping across Broad Street and starting new fires in another row of buildings. A resident rode horseback to Matawan, seeking help from its fire company; this is an early beginning to what is known today as mutual aid. The fire, alas, with all hands working feverishly, was winning. Suddenly the winds that had been driving the fire shifted and then died down completely.

On October 10, 1877, Keyport Fire Engine Company was organized. It incorporated on February 28, 1878, as the Keyport Hook and Ladder Company No. 1. On April 11, 1878, the town commissioners presented the first firefighting apparatus to this company, which had cost

$1,000. With no suitable building available, an arrangement was made with Rufus Ogden to construct a building for that purpose on Third and Broad Streets, a building also to be used by the town commissioners for their meetings. Since 1986, they have been located at 34 First Street in shared quarters with Keyport Engine Company No. 1.

The principal men who met to form an organized fire company included Thomas S. Rezau Brown (who chaired the meeting), Stephen V. Arrowsmith, Thomas Burrowes, F. D. Woodruff, James H. Johnson, Theodore Aumack, Albert Van Nostrand, Joseph T. Burrowes, George Holmes, Edward Bedle, C. H. Cook, Elias H. Force, Charles Winterton, William West, and William Van Dine.

Another group of about 15 men met in the Keyport Athletic Club on First Street. They decided to pursue a second company to assist in firefighting, one with "an engine of some description." The Keyport Engine Company No. 1 organized on August 5, 1889. It was successful with town officials and received a chemical engine in 1890. It was stored in the Keyport Athletic Club building on First Street until a new firehouse on Main Street was built. In 1889, Jacob Leyrer Jr. became the only firefighter to die in the line of duty thus far when he attempted to save his own father's life from their burning bakery.

Lincoln Hose Company No. 1 was the first protection afforded on the eastern locale of Keyport, referred to as Lockport. Organized on March 31, 1893, it was located on Waverly and First Streets behind the Central Railroad of New Jersey's freight yard. The first name chosen was the American Hose Company No. 1. Its first apparatus was a four-wheeled carriage, painted ornately in white and gold. In 1915, it became the first motorized fire company in town, using a secondhand Acme truck. In 1909, it became a combination hose and truck company, which is how it is still officially chartered. It is located at 142 Second Street.

Raritan Hose Company No. 2 became the fourth unit and was organized on May 29, 1893. Meeting in the armory building were Capt. Asbury Bedle, Alfred Cuttrell, and Richard White, who became the first foreman. With others and all the men of Company G Regiment, National Guard of New Jersey, they became charter members of the company. Originally, only military men were eligible for membership. They occupied two firehouses before moving into quarters on Division Street in 1900; this firehouse was moved to Broad Street in 1919 and is today the Keyport Fire Museum and Education Center. Raritan Hose moved into a new building at 1893 Maple Place in June 28, 1970, where it remains today.

Liberty Hose Company No. 3, organized on June 30, 1893, was formed in part by James Holmes and Alvarado Walling. A meeting was called by these young South Keyport men and held in Theodore Aumack's carriage factory, resulting in 14 charter members. Working on Labor Day, dedicated members carried lumber and material to save cartage cost to build their first home. Liberty Hose eventually became situated on Highway 36 to serve the southern section of town and remain in Mechanicsville to this day.

Eagle Hose Company No. 4 organized on June 17, 1907, after residents of Keyport's Brown's Point sought fire protection. Alfred Geran and others held the meeting in H. P. Disbrow's bottling works, resulting in charter members such as Grover and Winfield Bailey, Frank and Edward Cuttrell, Judson Hopla, and Maurice Maloney, to name a few. Eagle Hose Company No. 4 is today located at 95 Broadway in the town's oldest firehouse.

Keyport Fire Patrol, known as the Board of Fire Wardens, can trace its roots back to 1893, when the first fire ordinance created a fire police patrol. Some 10 members from each of the five companies united to form the unit. The 1912 ordinance established the unit as a separate company. In addition to being firefighters, this company's special duties include salvaging personal property, guarding fire scenes, and keeping the public away so as to ensure firefighters work uninterrupted. Keyport Fire Patrol is located on Waverly Street.

The bell has sounded! It is time to gear up, slide down the pole, and pull the fire wagon back through time as you enjoy *Keyport Firefighting.*

One

FIRES

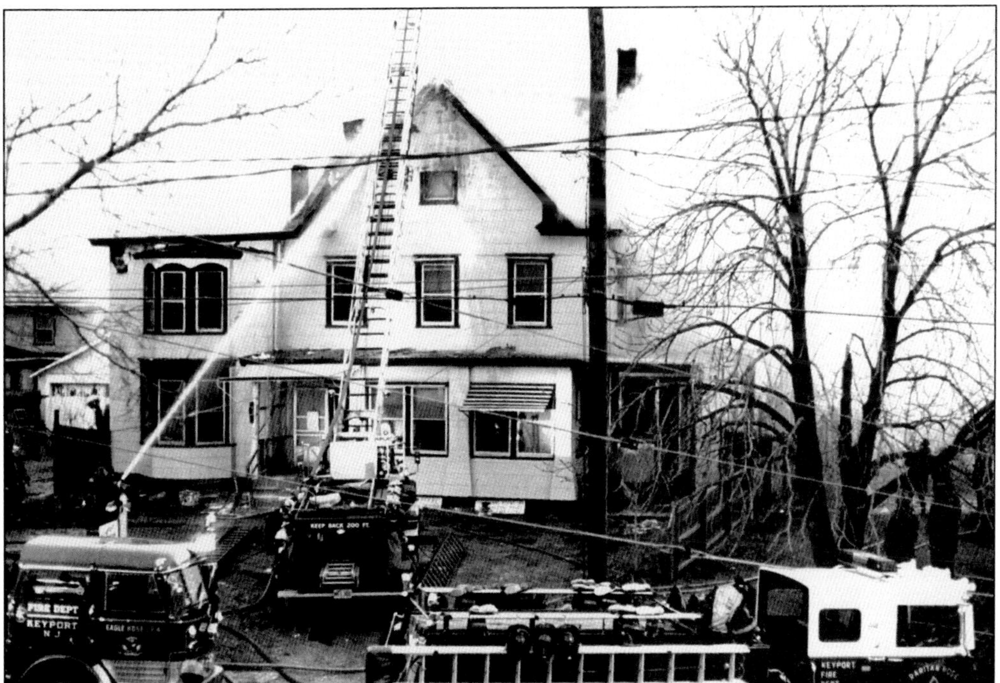

CHARTER HOUSE FIRE. The Brown's Point Tavern stood overlooking Raritan Bay for many years prior to the American Revolution. During its existence it operated as a Colonial tavern, stagecoach stop, and later a nursing home. It had escaped several previous fires, but in 1979, the building was consumed by a suspicious blaze and became just another memory.

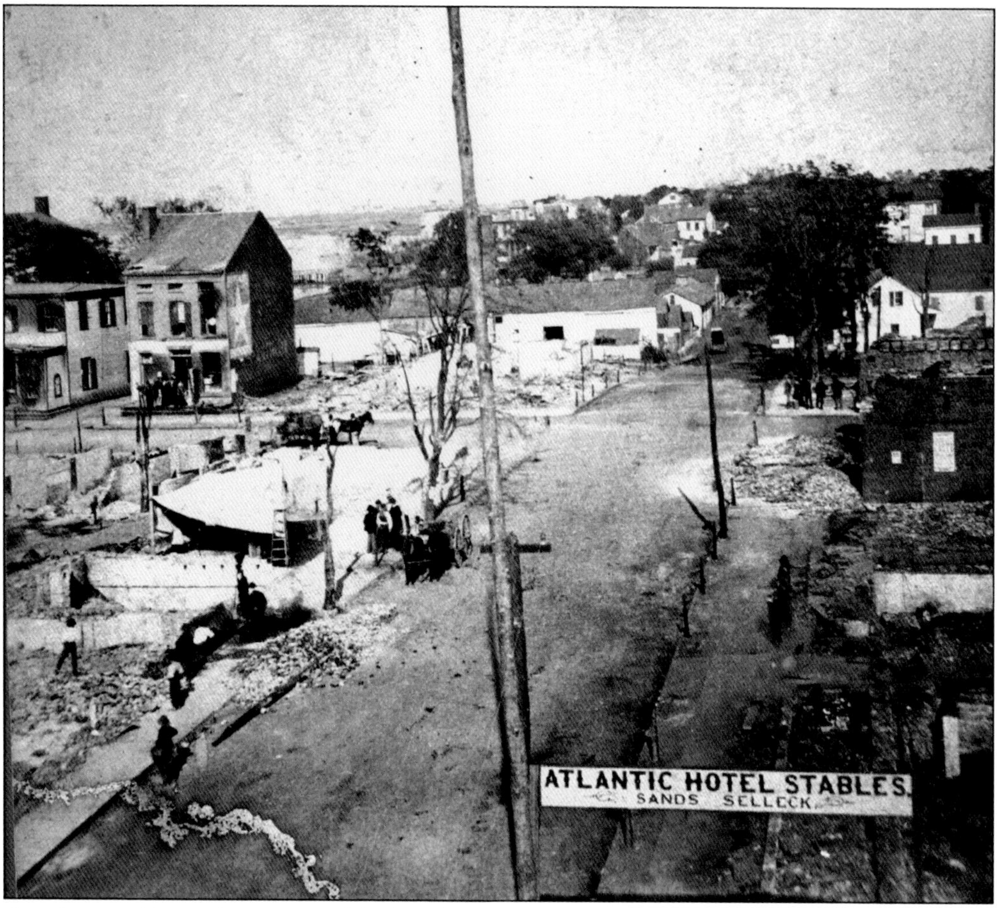

THE GREAT FIRE, SEPTEMBER 21, 1877. The town's merchants had been raising funds for about six months to purchase a steam-powered fire engine for the village. Then something fishy happened—someone made off with all the money. Within a few weeks of the theft, a small fire grew out of control and quickly consumed just about the entire business district. With a strong offshore wind blowing, it looked as though the town would be doomed. Building after building fell to the flames, including the Mansion House and Atlantic Hotels; soon the Episcopal church was burning from foundation to steeple. Within an hour, over 20 structures were burning out of control. Soon Matawan arrived with its hand engine and began valiant work in slowing down the spread of fire. Around 1:00 a.m., the wind shifted and slowed the spread of the blaze. By morning, Keyport's business center had been all but destroyed, and the people vowed that this would never again happen to them. On October 10, 1877, 100 men banded together and formed the Keyport Fire Engine Company No. 1, which was the predecessor of the Keyport Fire Department. This view looks east on West Front Street toward Broad Street.

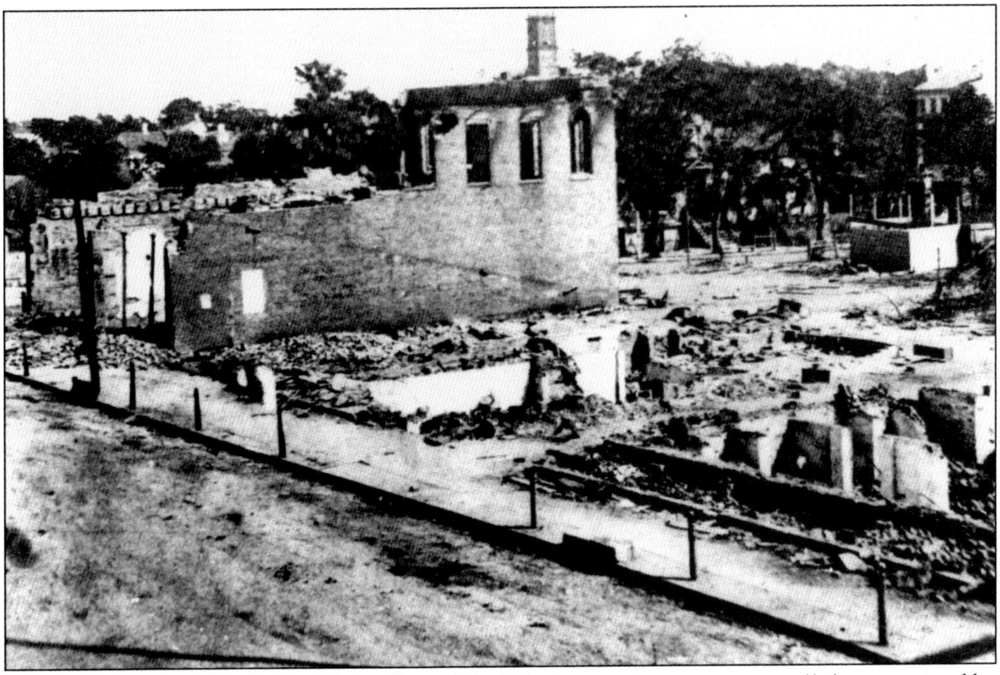

Great Fire, Looking East. The cellars of the West Front Street stores are all that remained by Saturday morning of September 22, 1877. In the background are the remains of Vice Pres. Garrett Hobart's childhood home, which stood on Broad Street.

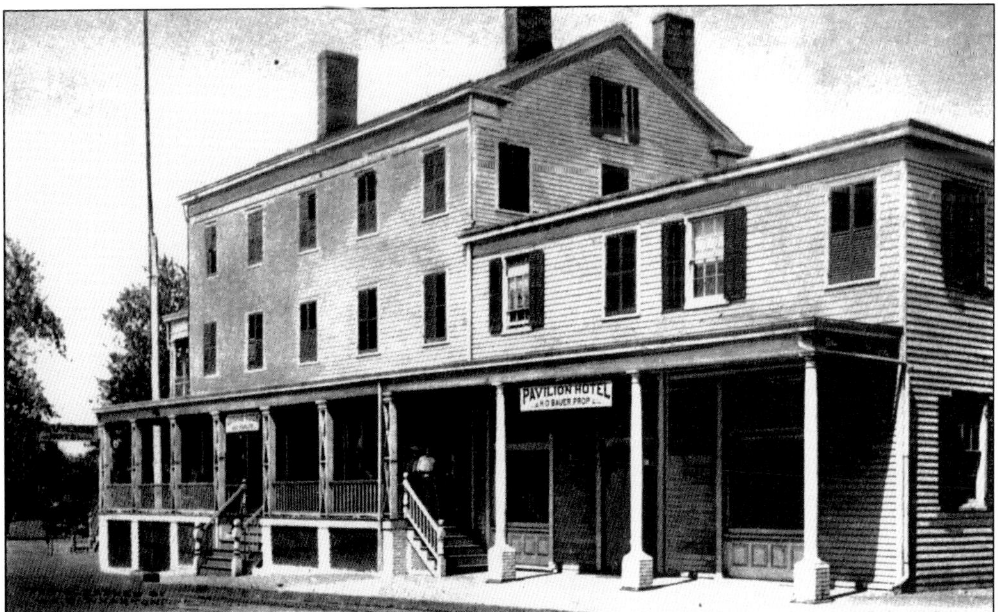

Pavilion Hotel, Built in 1851. The Pavilion Hotel was the only substantial building to survive the fire. It was only saved by means of a bucket brigade, which stretched from the street to the wharf, a distance of about 500 feet. Although scorched, Keyport's famous old hotel became the focal point of the rebirth of the town after the fire, only to be destroyed by fire in 1963.

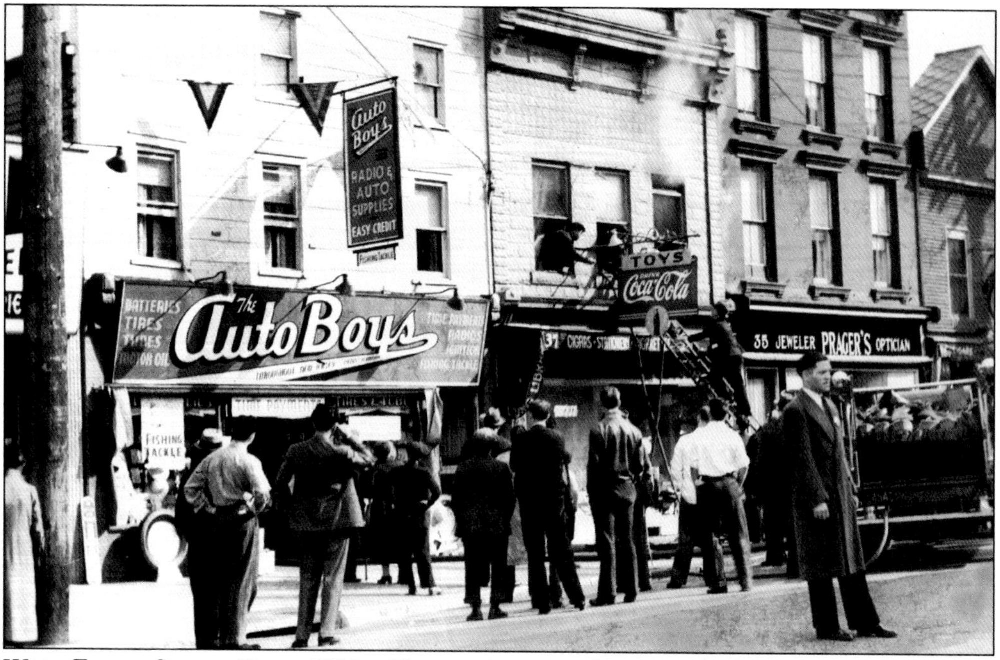

West Front Street Fire, 1930s. This quick-moving blaze nearly destroyed this building at 37 West Front Street had it not been for the quick action of the Keyport Fire Department (KFD). Ironically, the pumper of the Keyport Engine Company is parked directly at the spot where the wind shifted in the great fire, sparing the buildings in this view. (Photograph by James Lewis; courtesy KHS.)

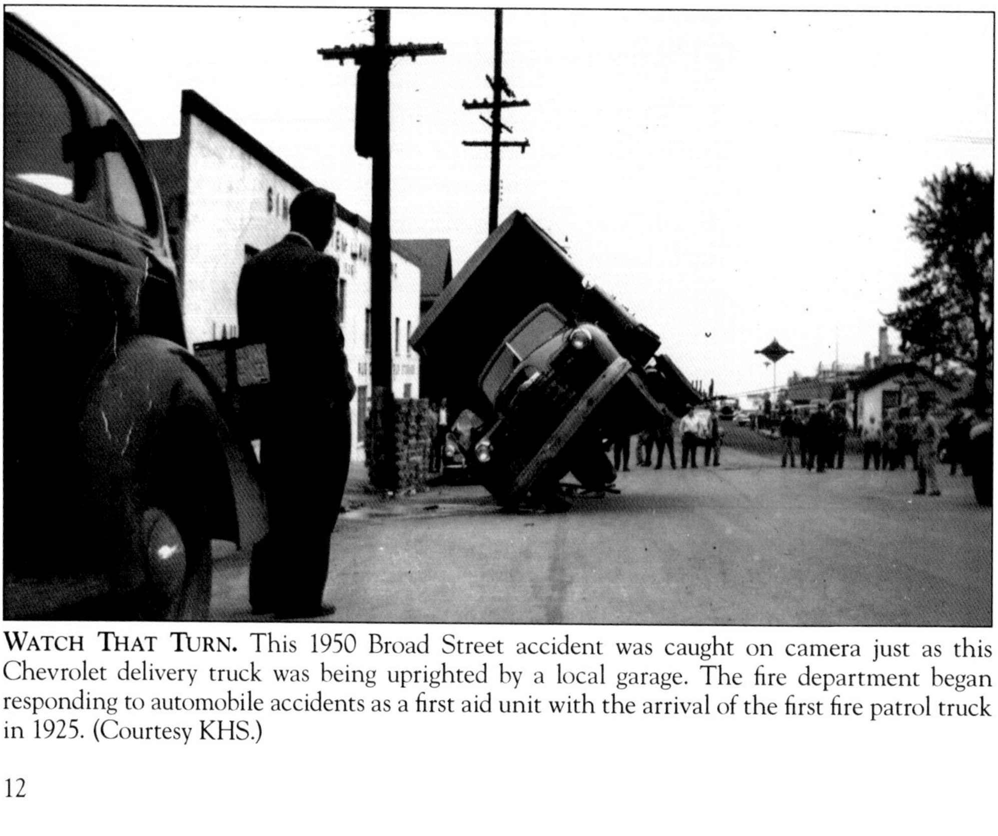

Watch That Turn. This 1950 Broad Street accident was caught on camera just as this Chevrolet delivery truck was being uprighted by a local garage. The fire department began responding to automobile accidents as a first aid unit with the arrival of the first fire patrol truck in 1925. (Courtesy KHS.)

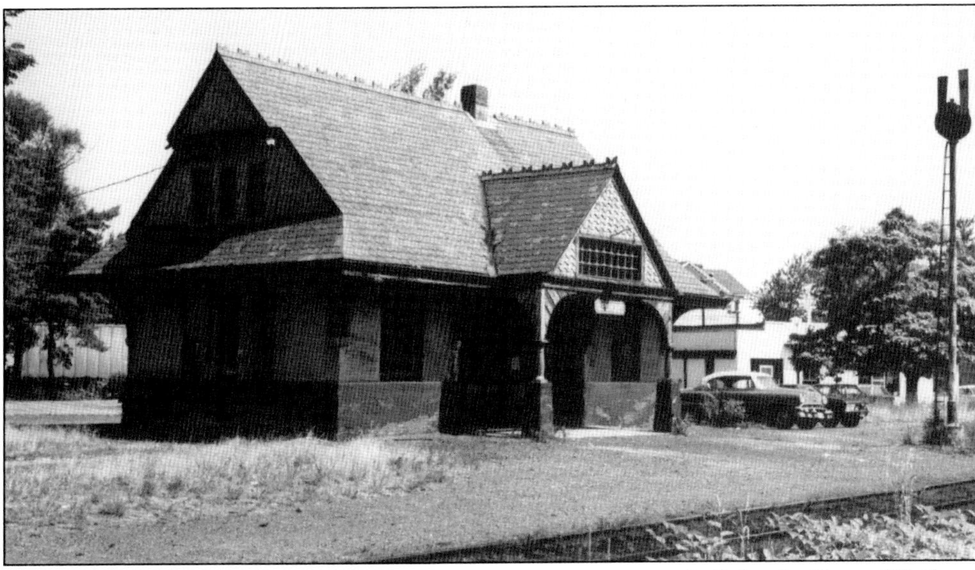

KEYPORT RAILROAD STATION. The Keyport passenger station was built by the Central Railroad of New Jersey in 1890 on the site of an earlier depot. This brick station was a one-of-a-kind design and was one of the more ornamental stations on the railroad's line from Matawan to Long Branch. Thousands of commuters, travelers, and soldiers all came and went from this quaint station, until it was closed in 1966 and passenger service ceased on the line. (Courtesy Tom Gallo.)

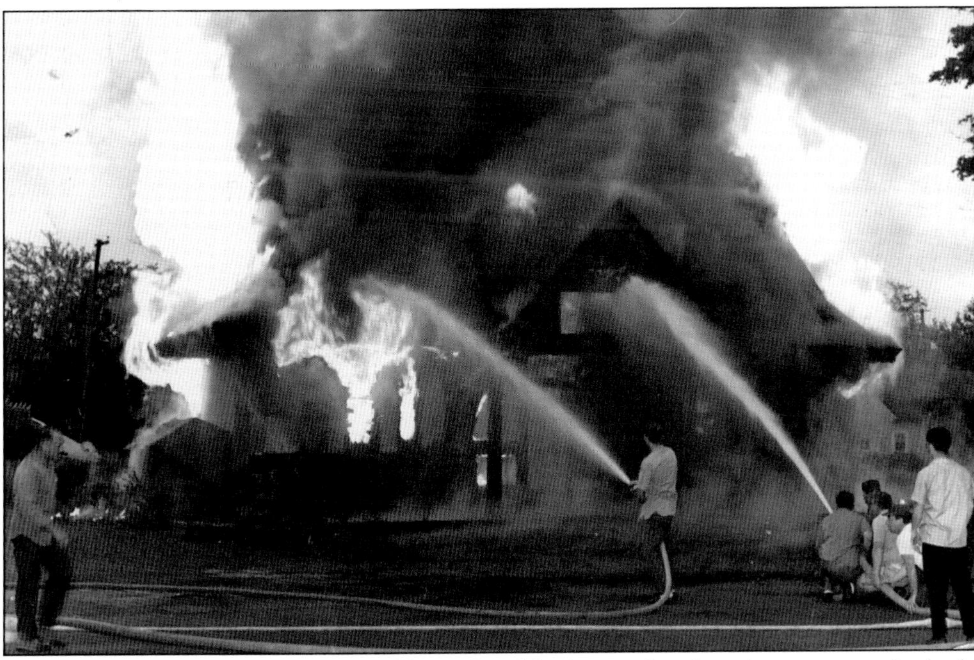

JULY 4, 1968. The railroad station had been closed for 21 months when this suspicious fire broke out early on this July 4 morning. A general alarm was sounded, and the entire department was soon hard at work to try to save the historic building. During this fire, firefighter Harold Dorl was caught between to fire trucks and was seriously injured. He later recovered and went on to serve as chief in 1971. Unfortunately, the station was not so lucky and was torn down a few days later.

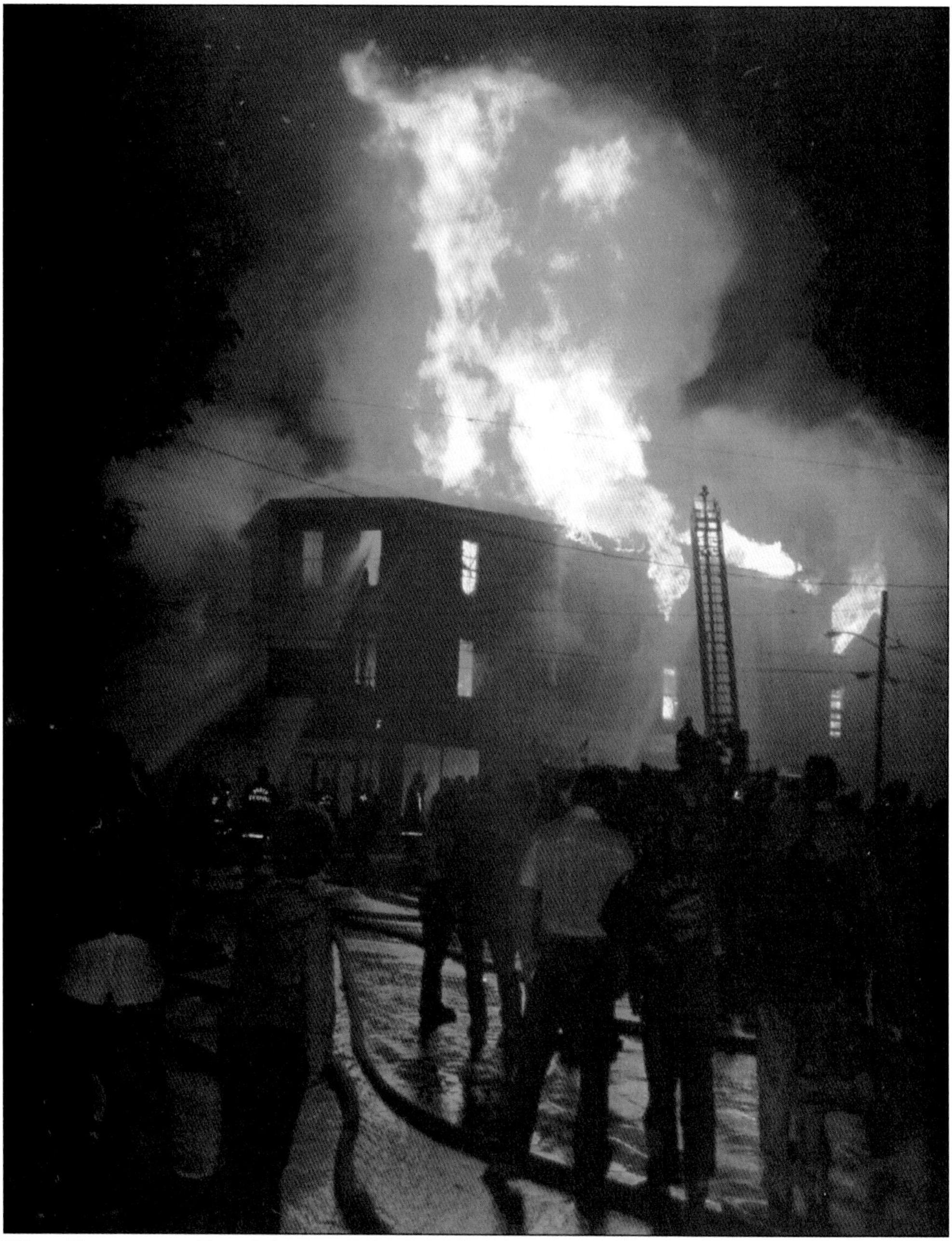

WEST FURNITURE FIRE. The worst fire to strike Keyport in over a decade occurred on Sunday, September 19, 1976. An arsonist's torch set the main showroom of the West Furniture Company ablaze around 11:00 p.m., just after most Keyport firefighters had returned home after attending the State Fireman's Convention in Wildwood. The fire, which brought about the largest request for mutual aid in the department's history, would have destroyed at least six other buildings if not for the valiant work of the department. One house owned by Al and Faye Bennett stood only 38 inches away from this fire and sustained only minor smoke damage throughout the home, which was built in 1848.

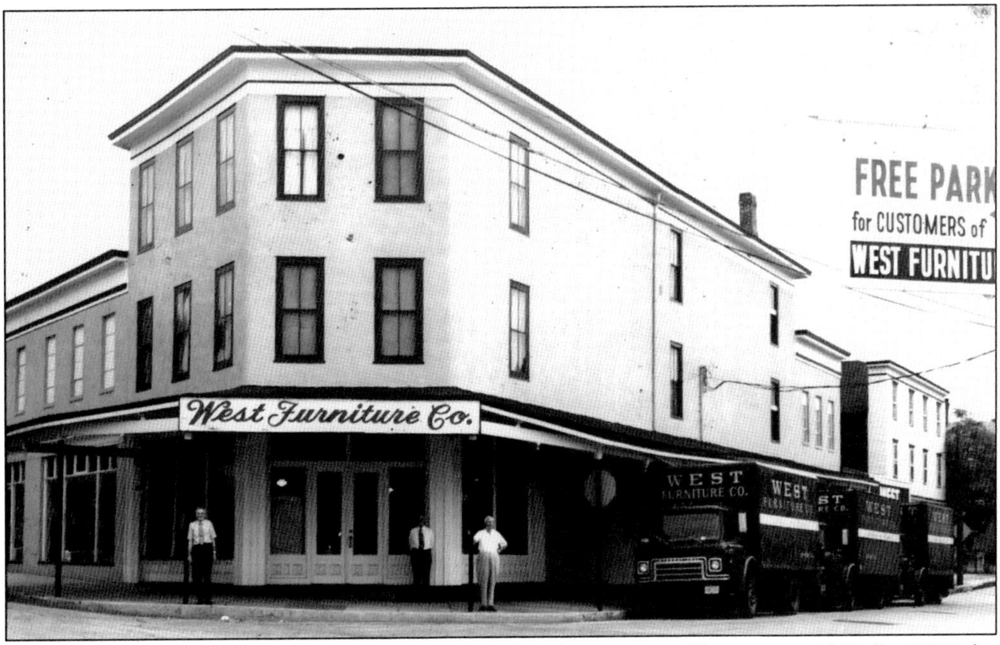

WEST FURNITURE, 1975. Robert West opened his first store in Keyport in 1869. By 1890, he had completed this three-story office and showroom on the corner of First and Church Streets. The building was one of the finest in the area and was filled with the best named furniture and carpets. By the 1970s, the West Complex consisted of a half dozen buildings around the old Church Street neighborhood.

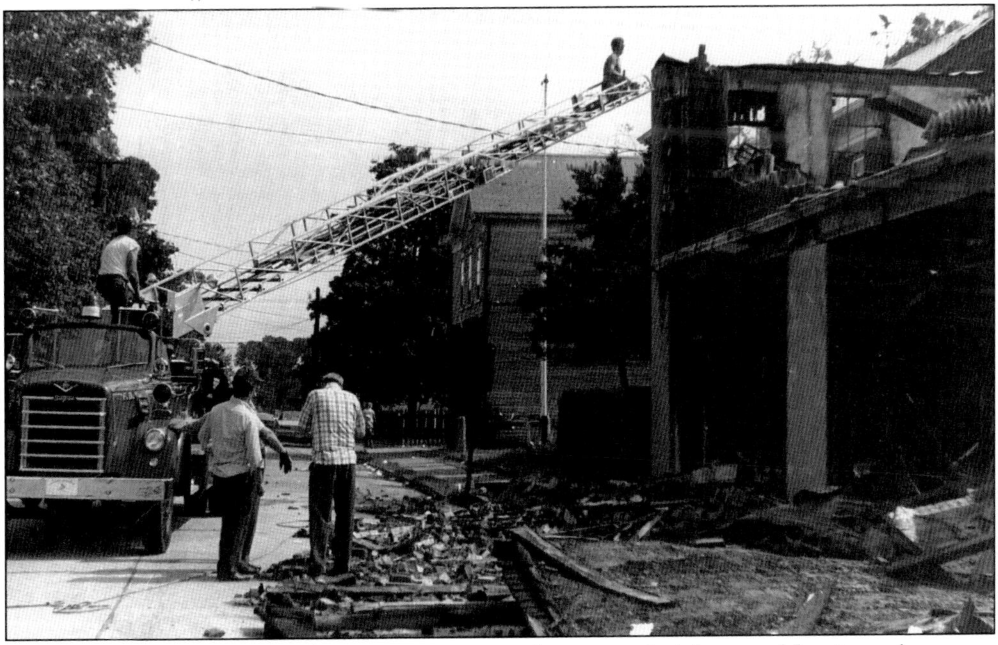

WEST AFTERMATH. By Monday morning, not much remained of the grand furniture showroom but a few walls and some chimneys; what had not been consumed by the fire collapsed into the basement and continued smoldering for several days. Fire companies worked in rotation to stay and wet down the hot spots for days. The building was never rebuilt.

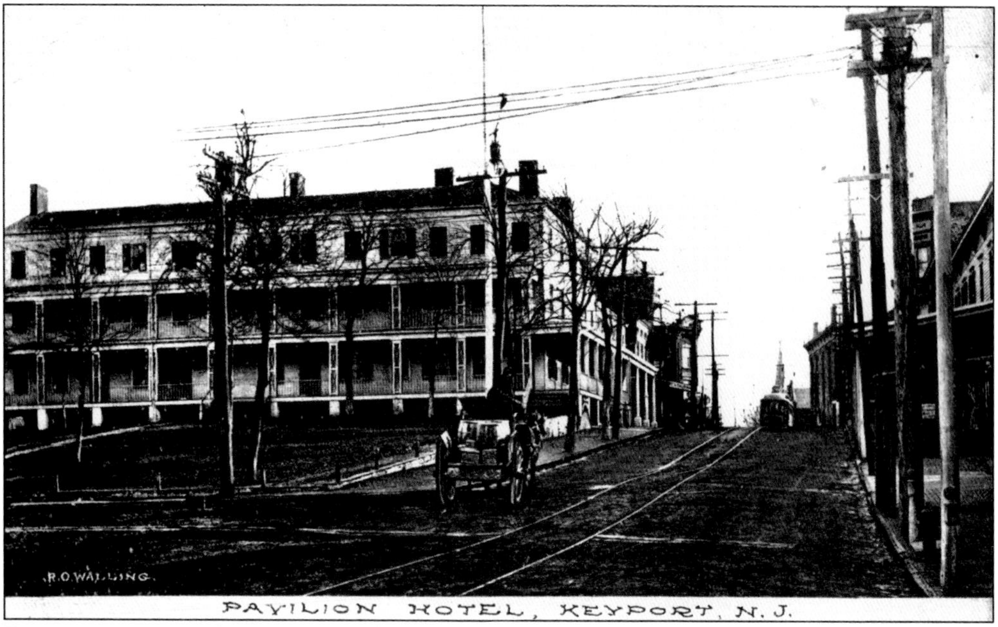

RARITAN INN. Northern Monmouth County's grandest hotel began life in 1841 as the Keyport Pavilion Hotel. Destroyed by fire 1851, it was rebuilt into the configuration seen here. Through the years, the old hotel that sat overlooking the steamboat wharf welcomed countless travelers, seamen, and merchants. By the early 20th century, it was renamed the Raritan Inn and sustained major damage and the loss of its third floor in a 1932 blaze.

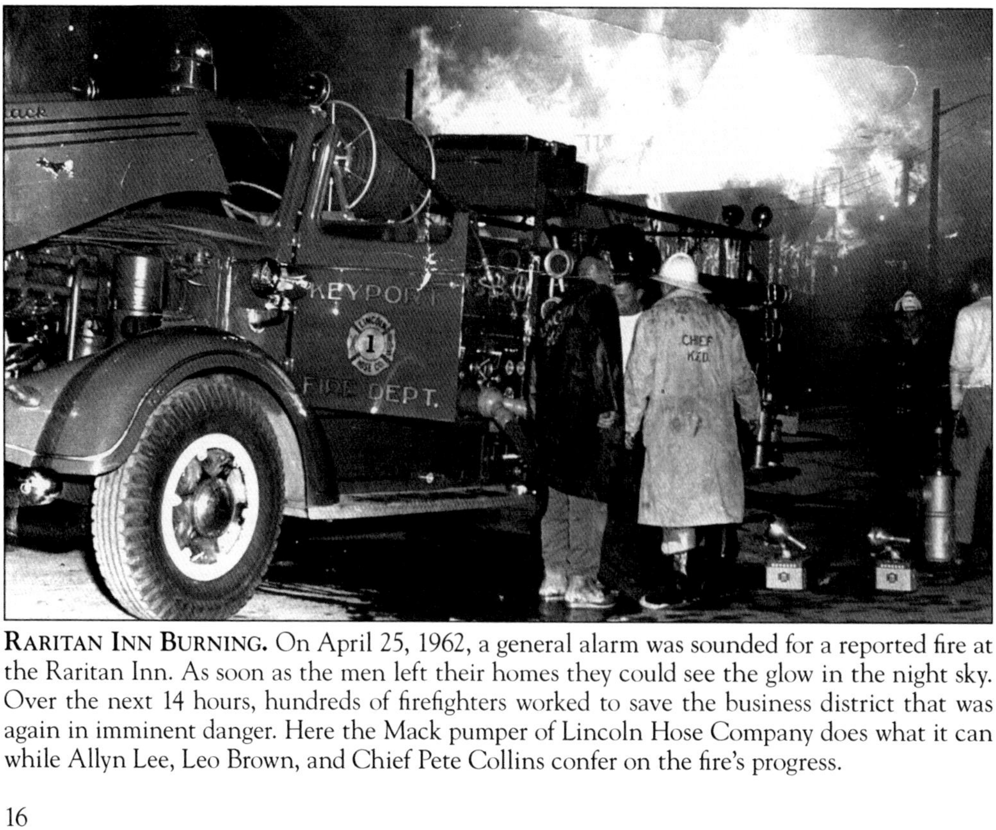

RARITAN INN BURNING. On April 25, 1962, a general alarm was sounded for a reported fire at the Raritan Inn. As soon as the men left their homes they could see the glow in the night sky. Over the next 14 hours, hundreds of firefighters worked to save the business district that was again in imminent danger. Here the Mack pumper of Lincoln Hose Company does what it can while Allyn Lee, Leo Brown, and Chief Pete Collins confer on the fire's progress.

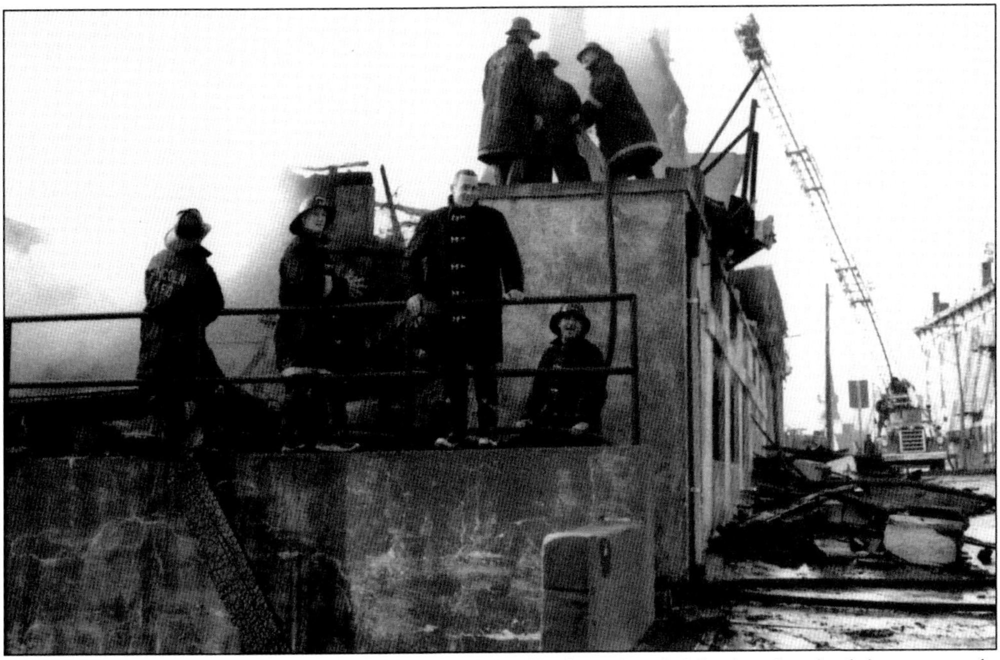

AFTERMATH OF THE RARITAN INN FIRE. By Friday, April 25, the fire had been mostly extinguished, and the danger to the town had been halted. Here Lincoln Hose Company members Jim Poling, Frank Van Pelt, and Dominick Panzera wet down the ruins of the grand old Keyport Pavilion Hotel.

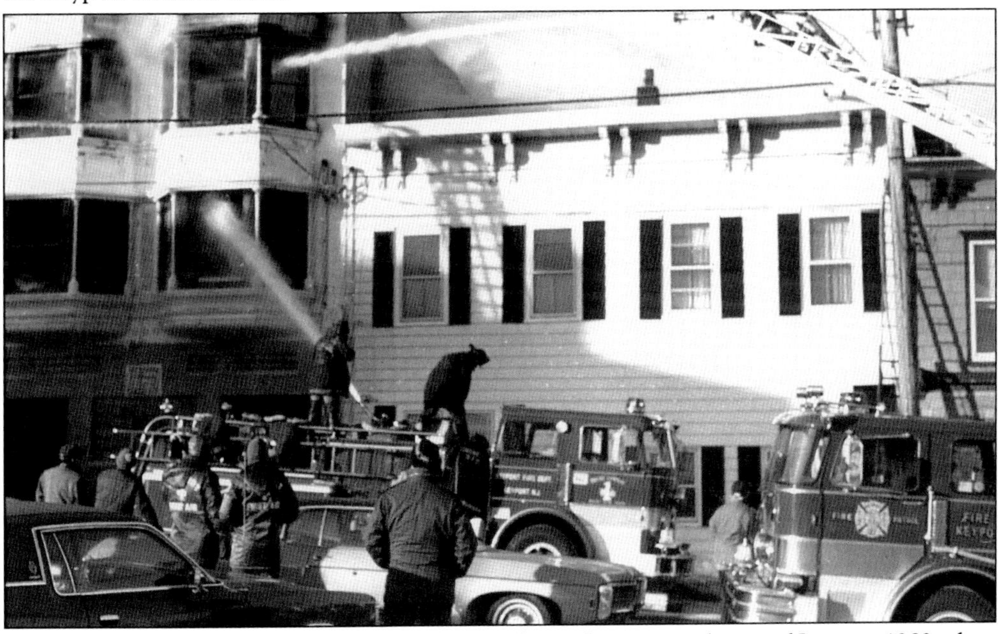

MONMOUTH HOTEL FIRE, 1982. It was during the early morning hours of January 1982 when this blaze broke out at the old Monmouth Hotel on Broad Street. This fast-moving fire soon engulfed the second and third floors of the structure, which had been built about 1888 by Luke Melle. During this fire, one of the hotel's residents was killed as he reentered the building to retrieve some money; he was not found for several days afterward.

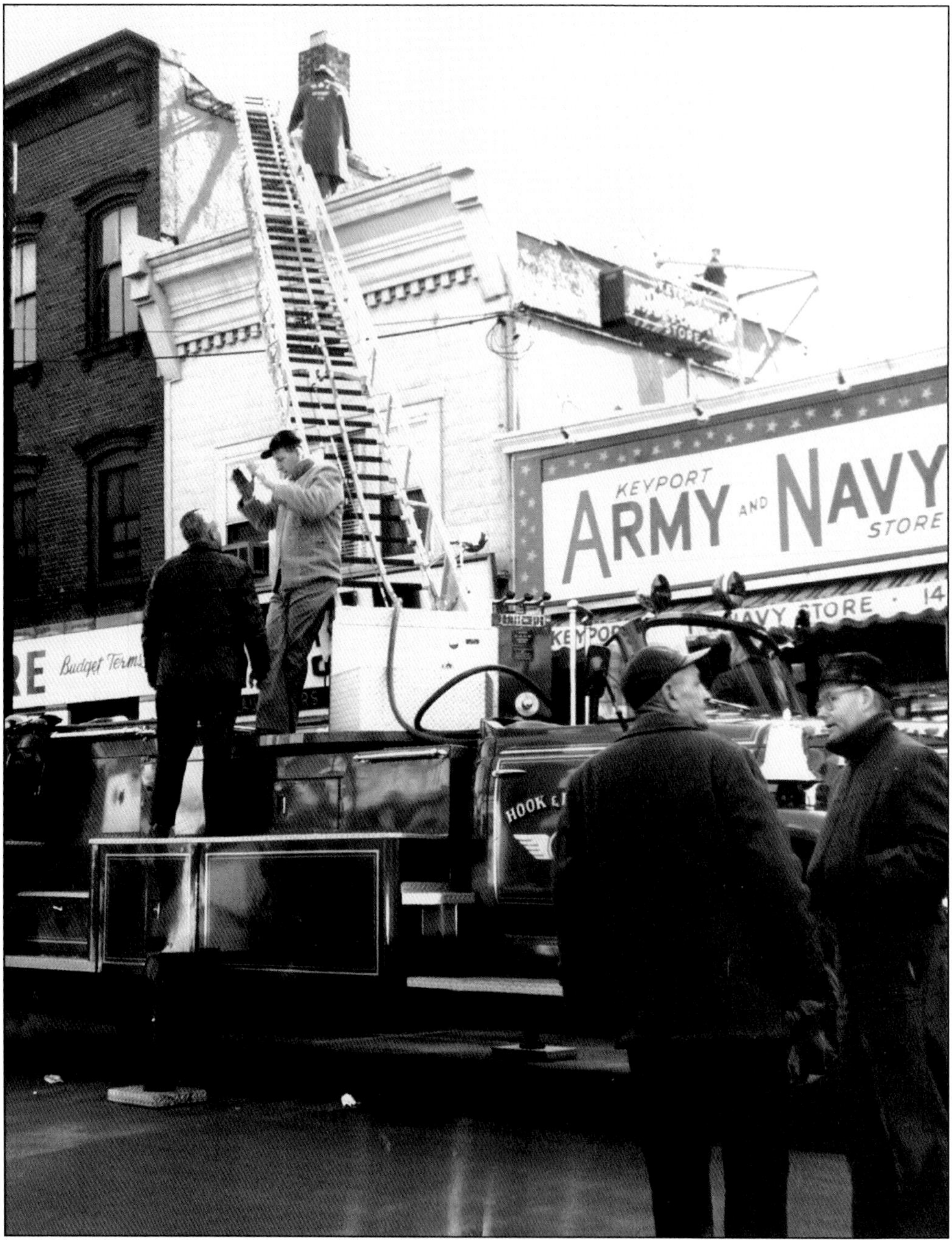

ATLANTIC AND PACIFIC STORE, 1961. Keyport's aerial ladder is used for two reasons, first as a ladder to access multistory structures and second as an elevated water source to force large amounts of water high up on burning structures. This day, the aerial ladder was needed for a small fire at the A&P store on West Front Street. Within a few years of this image, the 1878 stores were demolished for urban renewal.

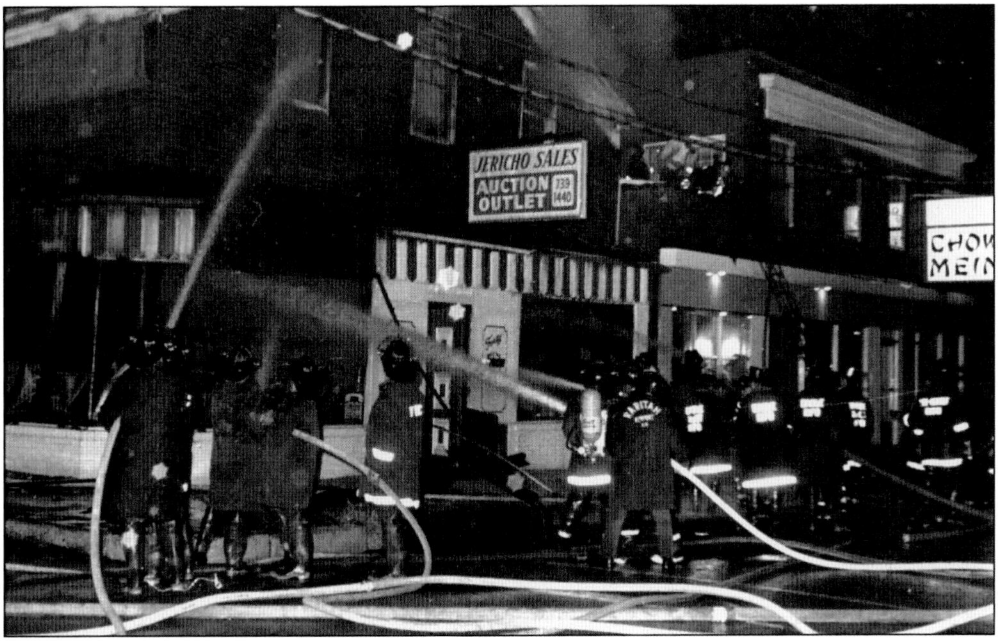

Jericho Sales Fire. Built as the Bedle Mill complex in the late 1840s, this group of buildings stood on the corner of Osborn and East Front Streets until November 26, 1973. It was located almost directly across the street from the old hook and ladder firehouse. This suspicious fire leveled the historic mill, and today the area serves as another downtown parking lot resulting from a fire. Below, members from Raritan, Eagle, and Keyport Engine Companies tend to the aftermath, ensuring all embers are doused.

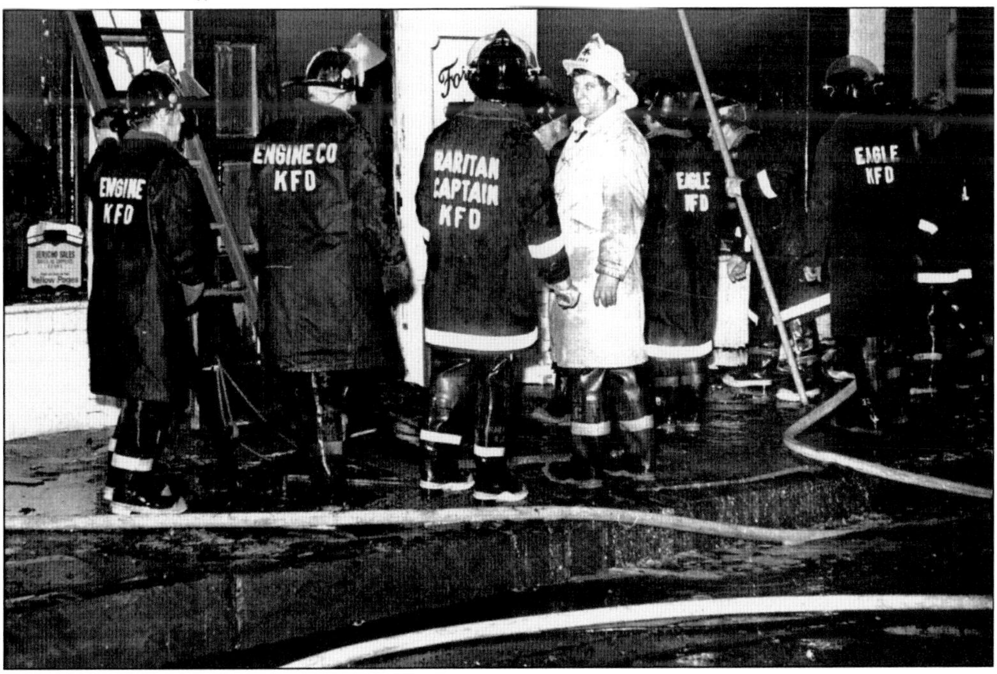

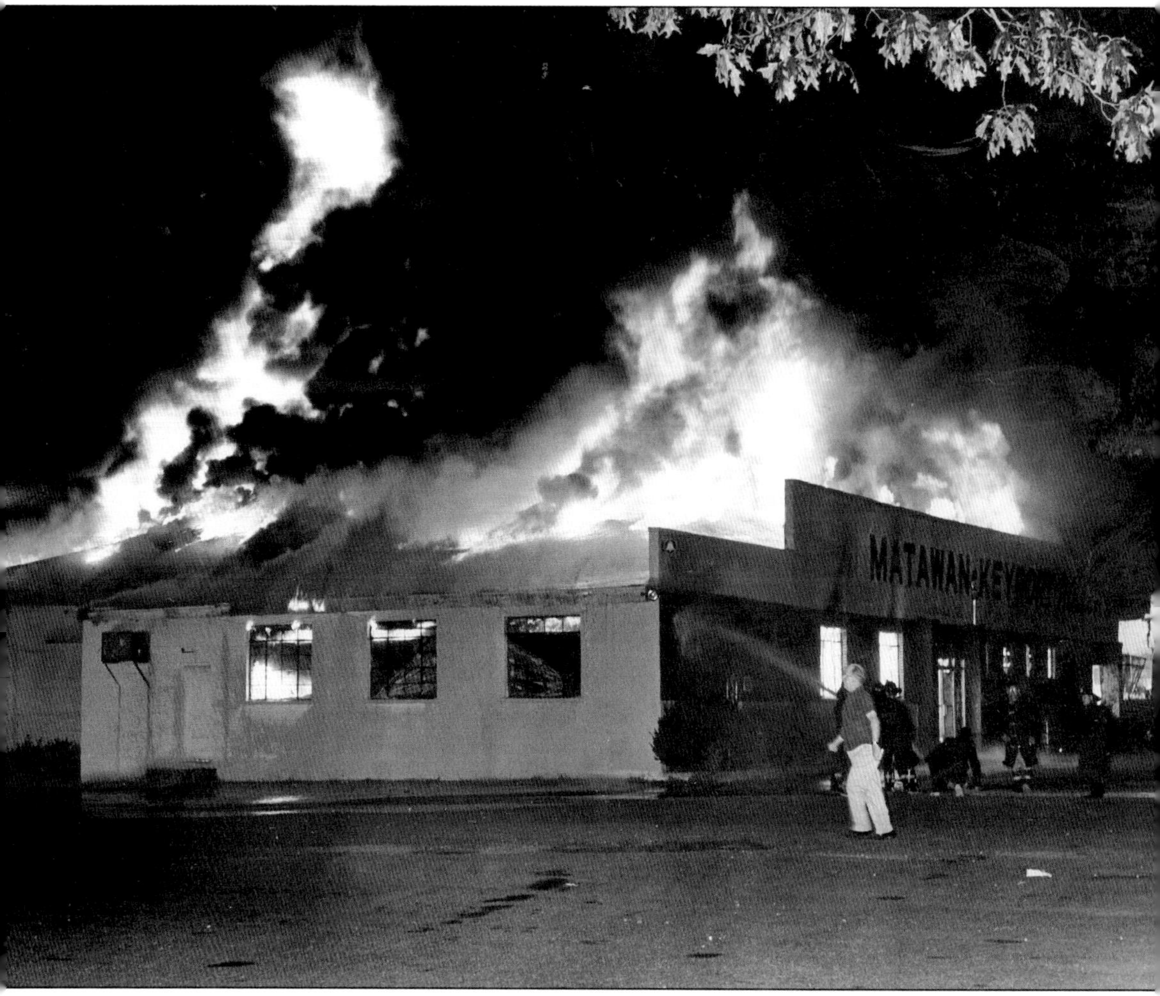

MATAWAN-KEYPORT ROLLER DROME. This large fire occurred in 1968 when a fire struck the local bowling alley and roller-skating rink. The building, which was built on the Keyport-Matawan (Aberdeen) Township border, was consumed quickly by a fire fed by the highly lacquered bowling floors. The building was not rebuilt, and today homes occupy the site. (Photograph by James Atkins.)

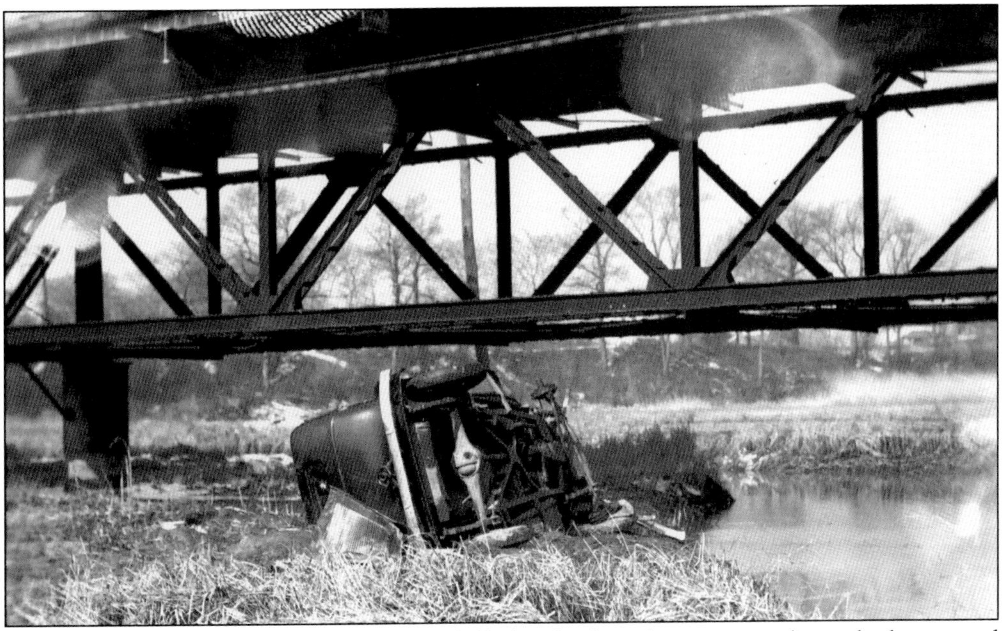

CAR OFF BRIDGE, C. 1928. One person was killed in this long-forgotten accident, which occurred when this car crashed through the guardrail of the Mott Street Bridge. The car nearly plunged into the creek below, instead landing in the marsh grass. Accidents such as this contributed to the fire department responding to incidents other than flames. (Photograph by James Lewis; courtesy KHS.)

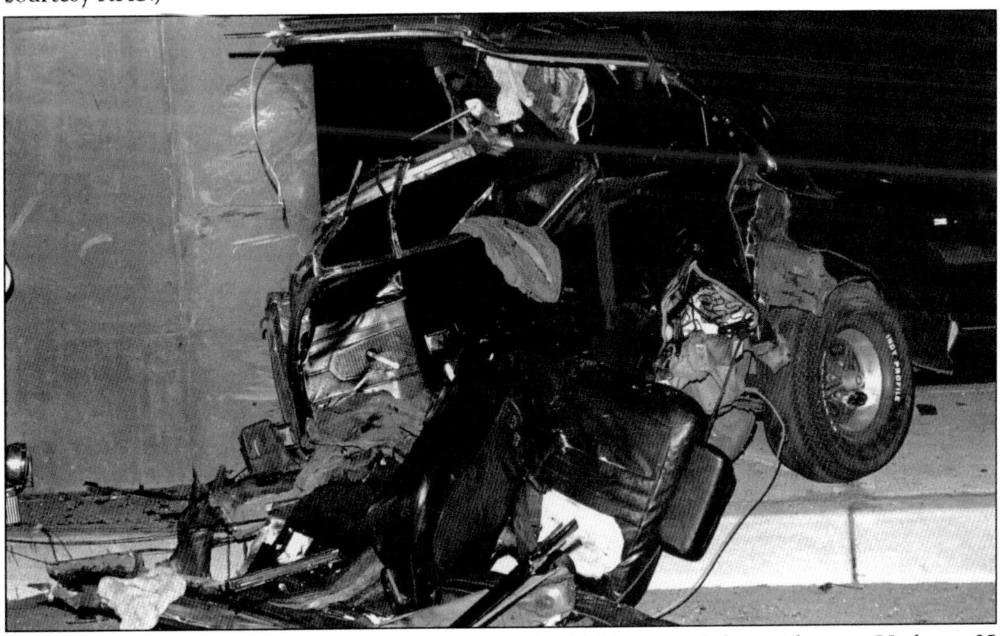

TRAGIC ACCIDENT, 1967. First arriving units to this 1967 automobile accident on Highway 35 were met with a grim scene. Four of six young local teenagers lay dead or dying in the remains of this horrific crash that occurred into the bridge pier of the Central Railroad of New Jersey. Obviously speed was a factor in this accident, which remains the worst single-vehicle accident in Keyport history.

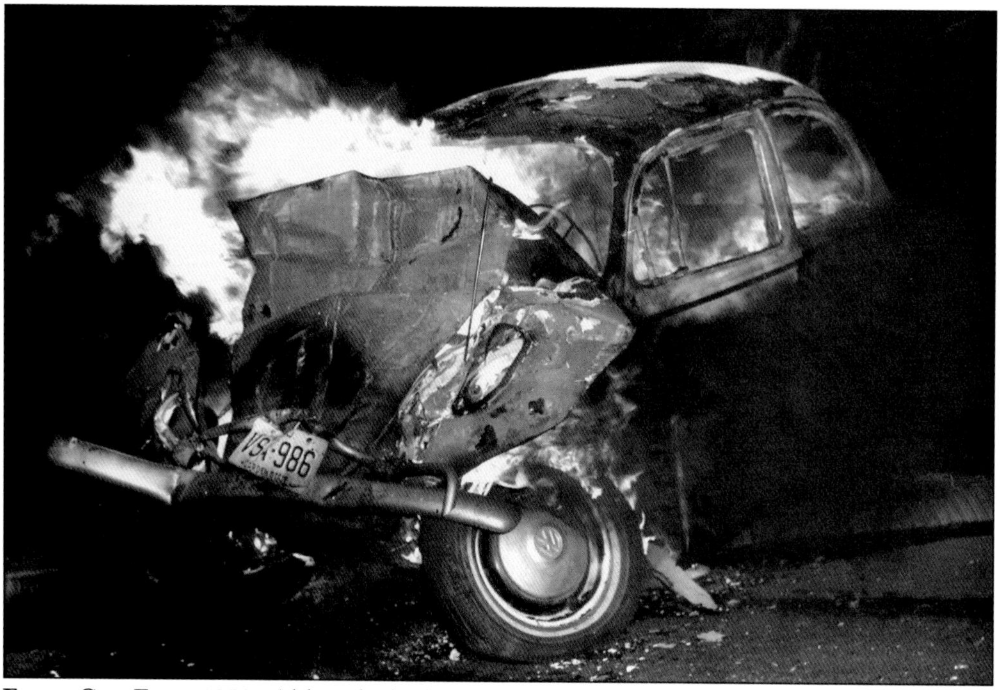

FATAL CAR FIRE, 1972. Although the borough of Keyport has only about two miles of state highways, they are among the most deadly stretches of state roads. Speeding vehicles often cannot properly negotiate the vertical and horizontal curves installed in the late 1920s to pass under the railroad as part of New Jersey's growing highway system. In this accident and a subsequent fire, a Hazlet man lost his life along Route 35, becoming the first Keyport fire fatality of the 1970s.

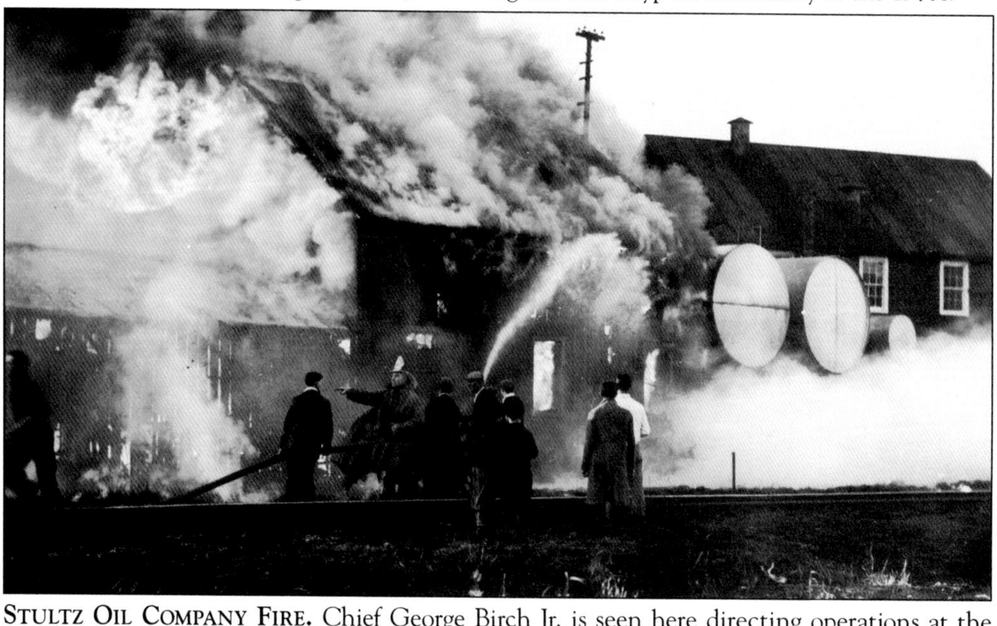

STULTZ OIL COMPANY FIRE. Chief George Birch Jr. is seen here directing operations at the Stultz fire of 1937 along the railroad tracks near Broad Street. Keyport firefighters prevented the large fuel oil tanks from burning during the blaze, but the outbuildings were a total loss. (Photograph by James Lewis; courtesy KHS.)

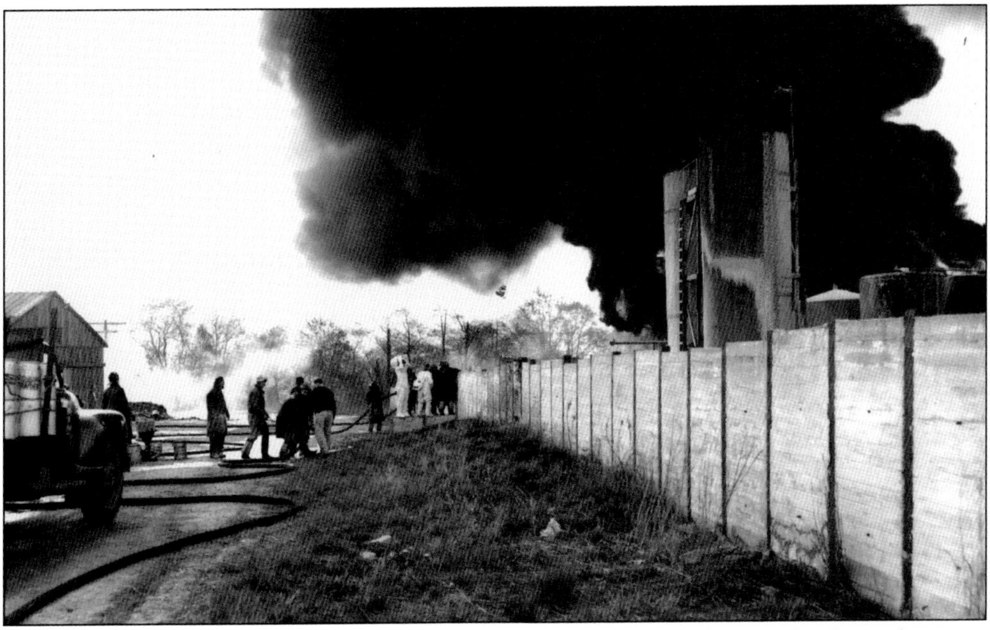

Chemical Plant Fire, c. 1952. The Nord Chemical Plant of Clark Street has been the scene of several serious fires over the years. This one broke out in the early 1950s, and soon firefighters from Keyport were joined by companies from all over the Bayshore area. Look closely and firefighters can be seen wearing two asbestos fire entry suits, one of which is today on display at the Keyport Fire Museum and Education Center (KFMA).

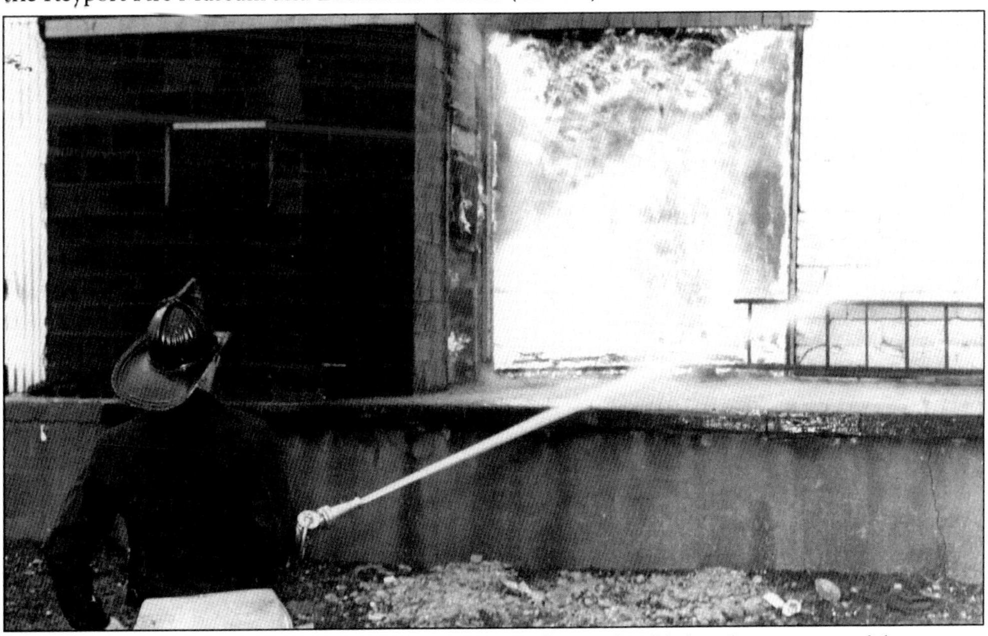

Conover Fire, c. 1955. This daytime fire has gained considerable headway in one of the storage buildings shown above. Here an unknown old-timer is wearing his 1930s vintage aluminum fire helmet, and he sees no need for a fire coat since it was "not cold enough to wear one." Today all firefighters must be in full protective clothing to operate at any incident to keep them safe and unharmed. (Courtesy James Atkins.)

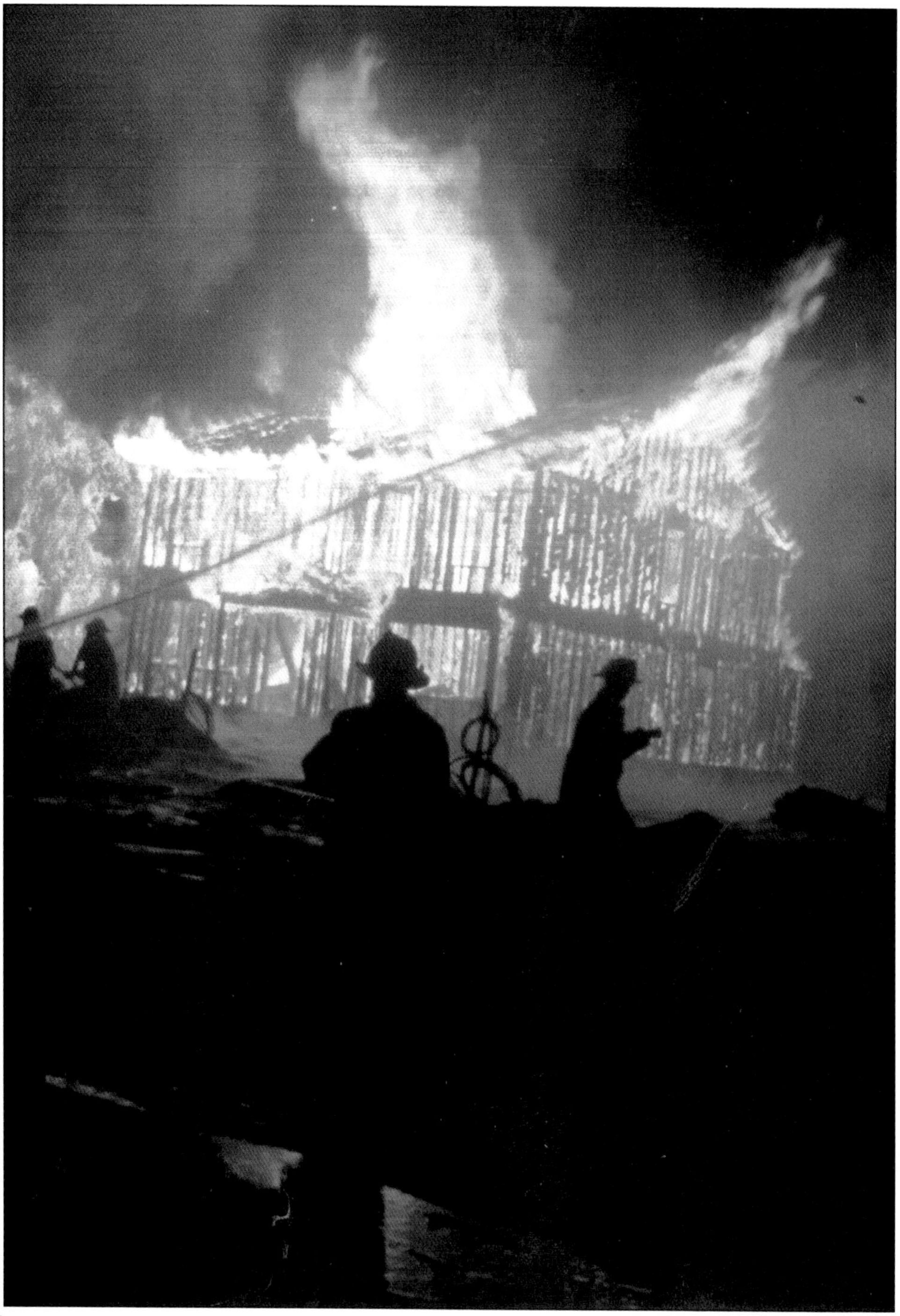

Coluco Place Fire. Firefighters are seen here silhouetted by the massive display of flames that devoured this house under construction in 1985. Chief Bob Poling was first on scene and directed the operations of all Keyport firefighters to help contain this blaze to the original structure. These buildings were constructed on the land known as Conover Field that was adjacent to their storage yard facility in the preceding photographs. (Courtesy Keyport Fire Patrol.)

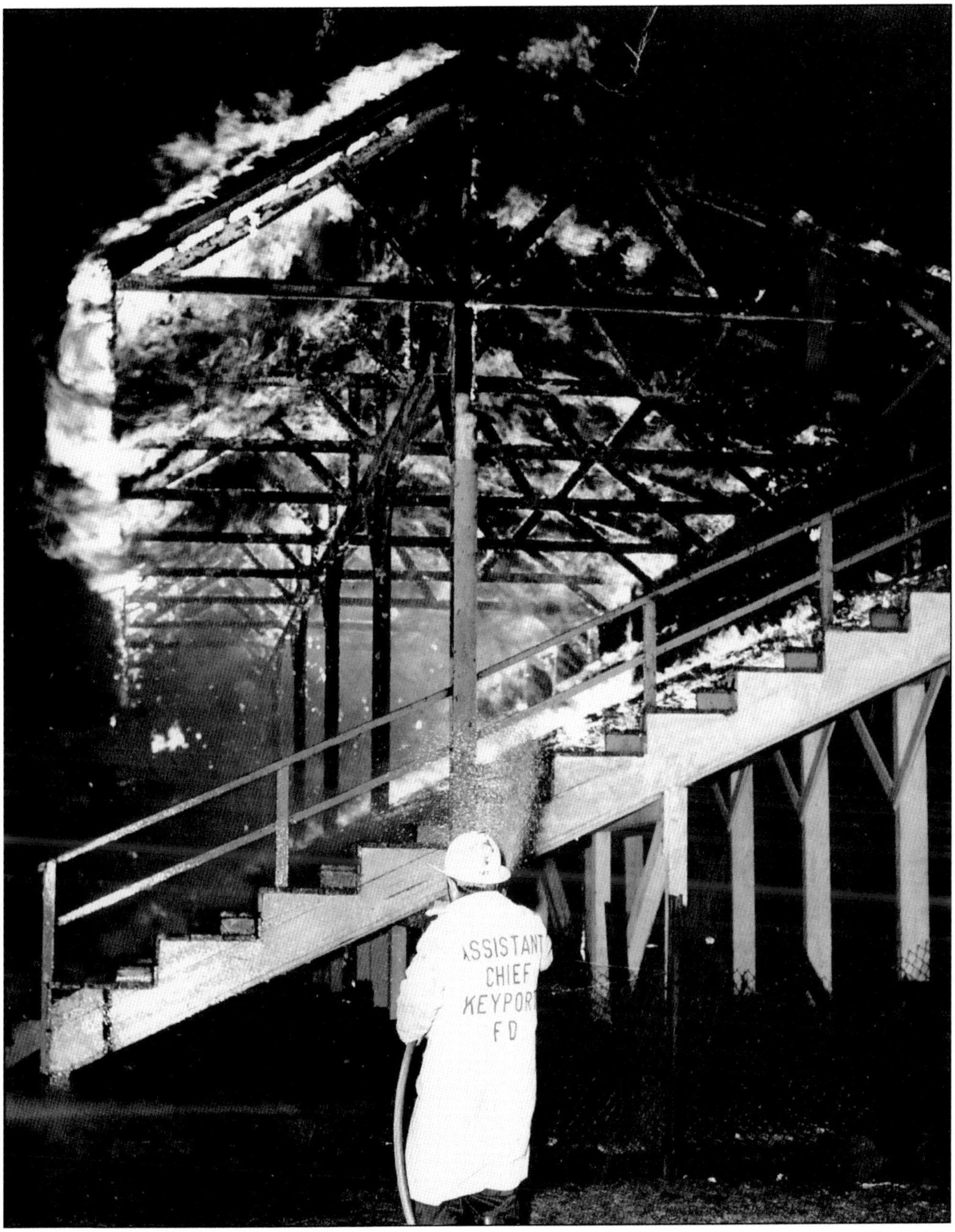

HIGH SCHOOL GRANDSTAND FIRE. Completed by the Works Project Administration in 1939, the baseball field grandstand met a fiery demise in 1970. Here an assistant chief stretches the first booster line until the larger hose lines can be put into operation by the next arriving companies. A booster line is capable of flowing about 30 gallons of water per minute while a two-and-a-half-inch line can flow upward of 200 gallons per minute. (Photograph by James Atkins.)

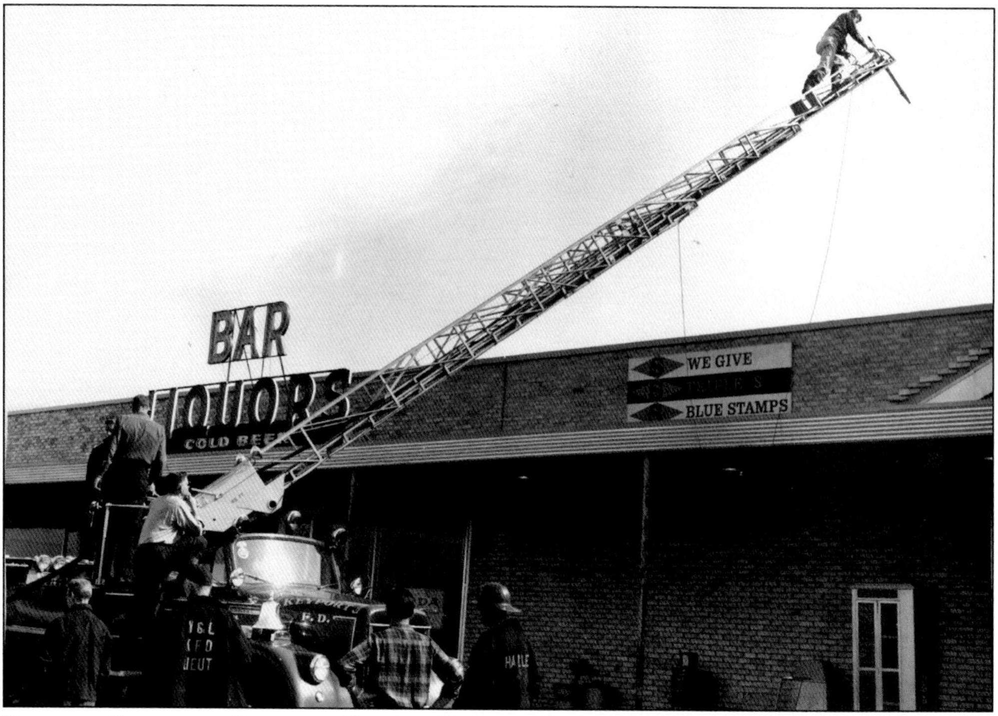

KEYPORT SHOPPING CENTER. The first Keyport Shopping Center was completed along the Route 35 and Route 36 corridor in the mid-1950s. The buildings were heavily damaged in a 1960s tornado.

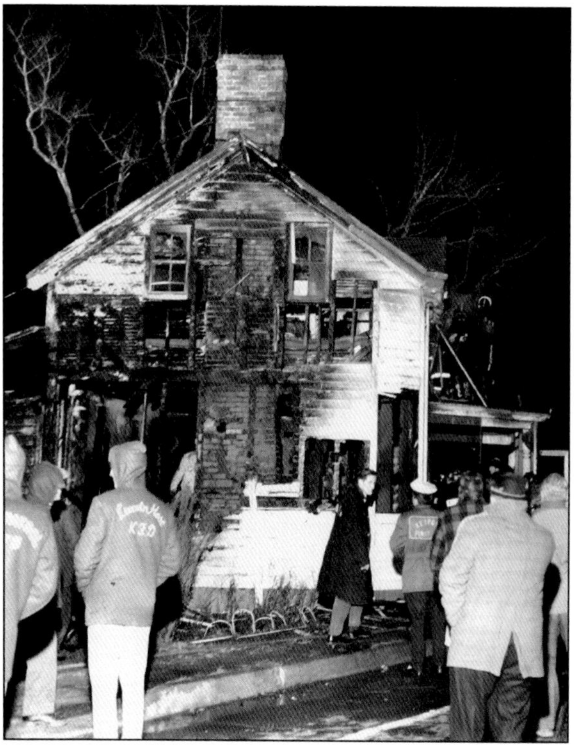

DAVIS HOME BLAZE. This old farmhouse on Route 35 near the Hazlet Township border was destroyed by fire about 1955. One of the oldest structures in the area, it was built using a balloon-type frame construction, which allowed the fire to travel from the basement to the attic in hollow voids within the walls. The house was completely destroyed and replaced by a combination dwelling and commercial building, which has been the scene of several small fires over the years.

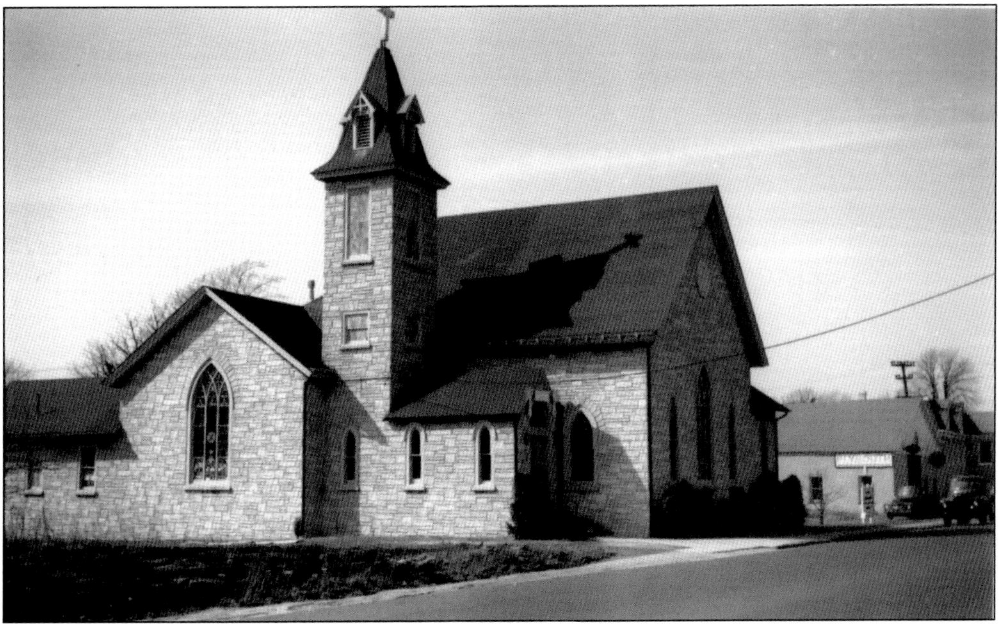

ST. JOHN'S UNITED METHODIST CHURCH. Built in stages from 1821 to 1930, the St. John's United Methodist Church was originally located on Holmdel Road near Bethany Road. In 1869, the original building (the part with the large stained-glass window) was moved to the Keyport-Hazlet border on South Main Street. It was later brick faced with artificial stone and was nearly 150 years old when disaster struck.

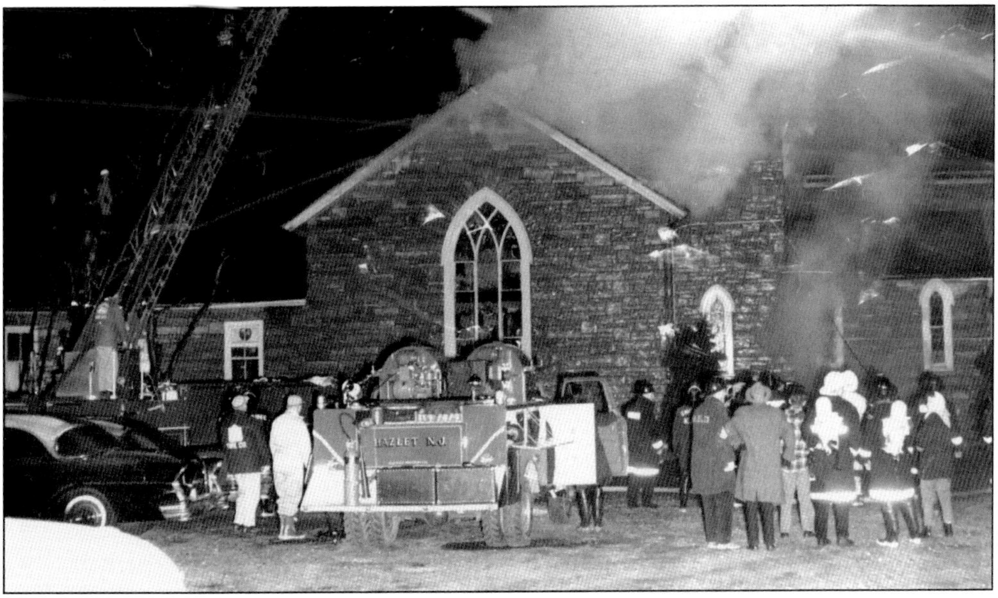

ST. JOHN'S CHURCH FIRE. In 1962, the Hazlet and Keyport Fire Departments responded together for a report of a fire at the church. Within a few hours, several hundred firefighters from all over the Bayshore were on scene battling the blaze. The fire burned fiercely in the attic and bell tower, and it was soon realized to be a total loss. The church was demolished, and the property is now part of the Remsen Dodge parking area and State of New Jersey jug handle for Route 35 and Holmdel Road.

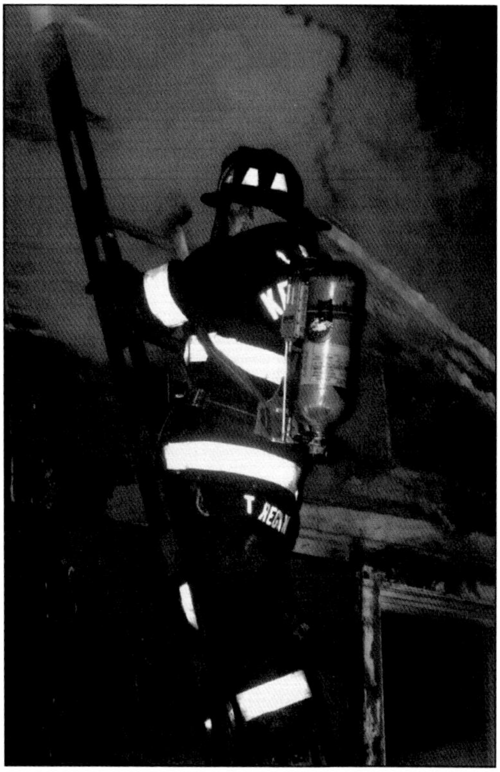

Time to Ventilate. In any structural fire, releasing hot gases and smoke from the building accomplishes several things. First, it reduces the interior temperature and exhausts the smoke so that firefighters can find victims and the fire quickly. Second, it controls the spread of the fire by allowing fire to seek oxygen as a fuel and travel out of the structure and away from the contents. Here coauthor Tim Regan ascends a ladder with an axe to ventilate an attic window.

Hose Line Management. In any structural fire, the hoses must be "stretched" first to the building, then into the building, and finally to the seat of the fire. Without utilizing proper techniques, hoses can become kinked and firefighters could lose pressure, endangering themselves. In this view, firefighter Scott Eack of Liberty Hose Company brings an inch-and-three-quarter hose line into a blaze.

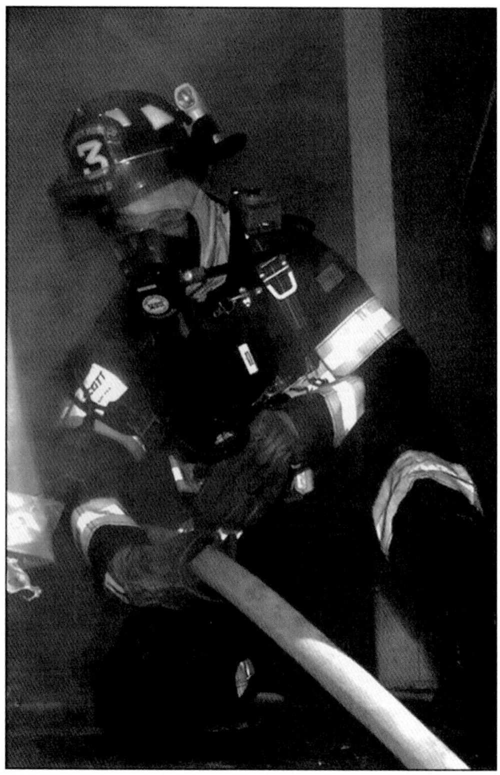

Two

MEMBERS

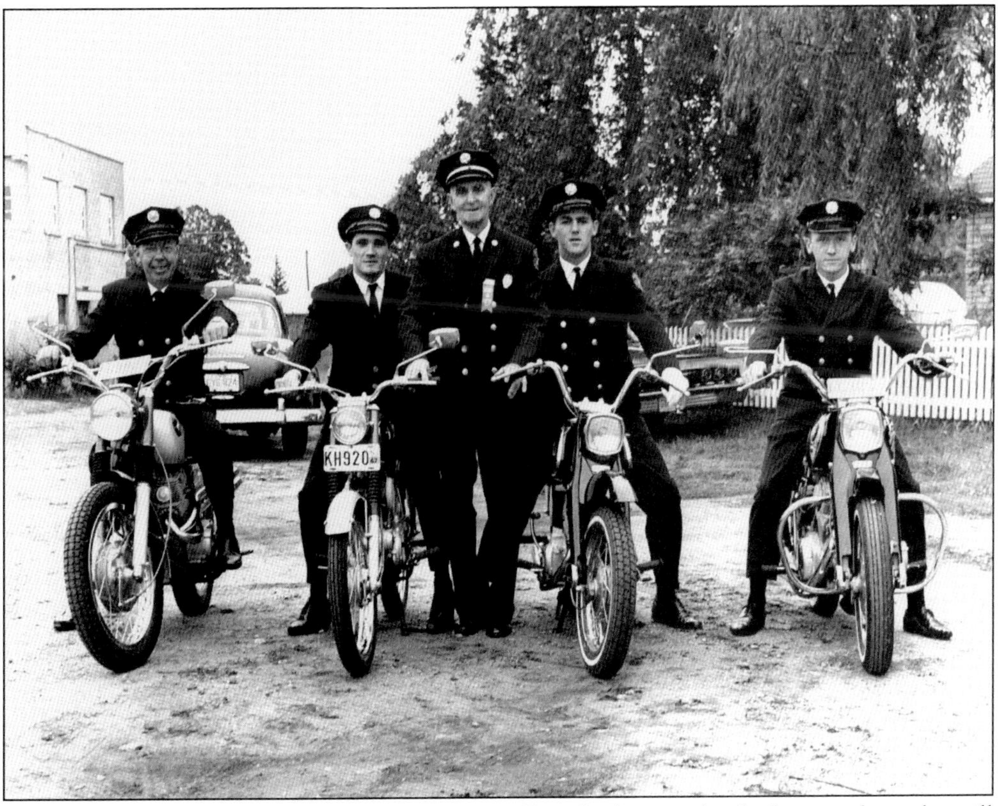

THE DEPARTMENT'S 90TH ANNIVERSARY, 1967. Everybody is set for the big parade, and it will not be long before the marching begins. Here Keyport firefighters, from left to right, Meldon "Babe" Stultz, Pete Rumsey, grand marshal Harry Poling, Terry Hicks, and Ken Young get ready for the big parade celebrating the department's 90th birthday party on October 4, 1967. (Courtesy Eagle Hose Company.)

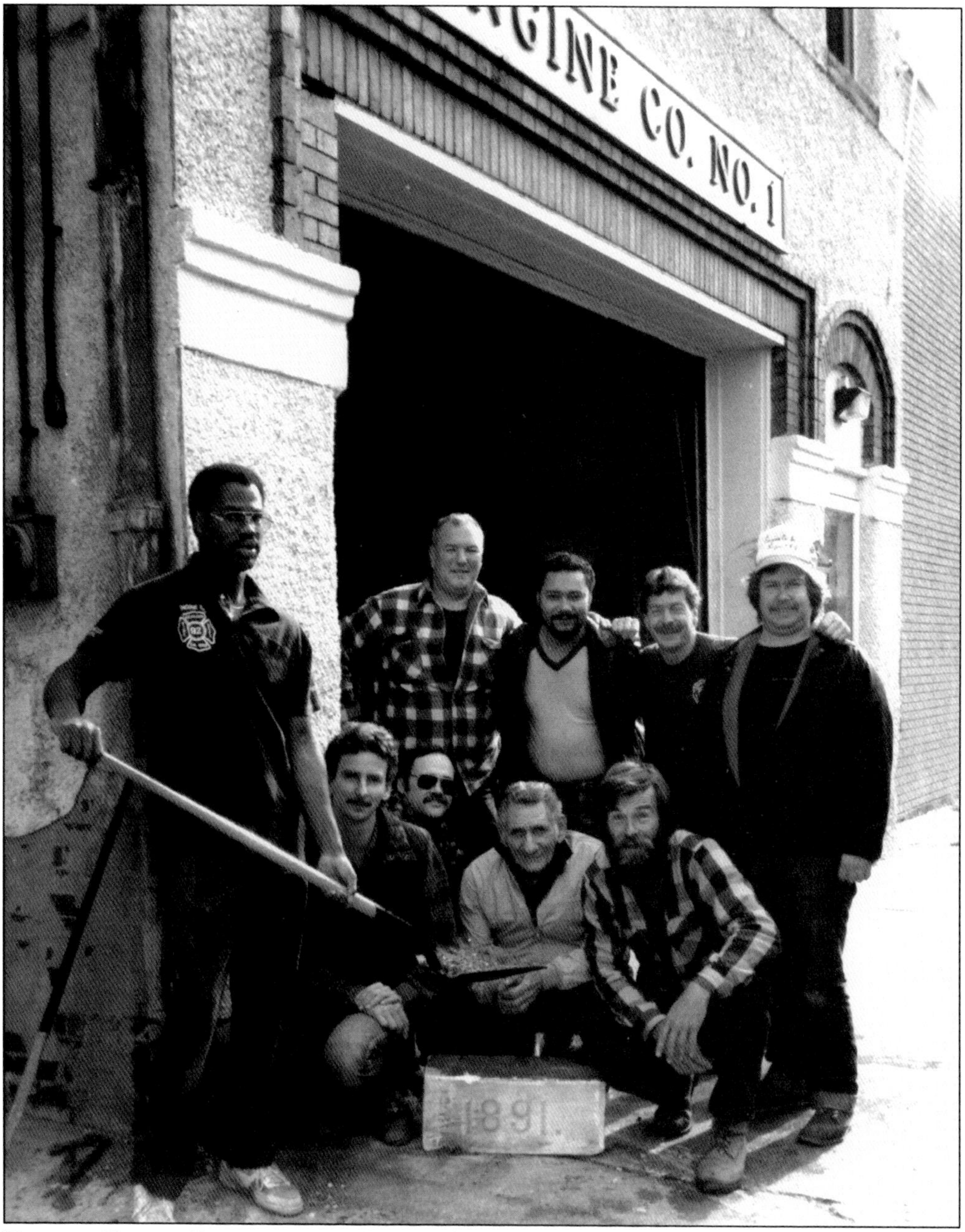

KEYPORT ENGINE COMPANY, 1985. During the fall of 1985, Keyport Engine Company members removed the 1891 cornerstone to locate the time capsule preserved within. Present that day were, from left to right, (standing) Jim Lawson, John Brandauer (chief, 1988), Jose Medina, Meldon Stultz Jr., and Ray Hedglin (chief, 1995); (kneeling) Kevin Morse, Bobby Baker, Jack Ross, and Russ Loshen. In just a few months, the company would relocate to new quarters at 34 First Street.

LIEUTENANT QUINN. David Quinn joined the KFD in 1997. He is the grandson of George "Buck" Leone, who served as chief of the department in 1961. Lieutenant Quinn was one on many Keyport firefighters who responded and assisted in the rescue recovery operations during the World Trade Center attacks on September 11, 2001. (Courtesy Pete Massas.)

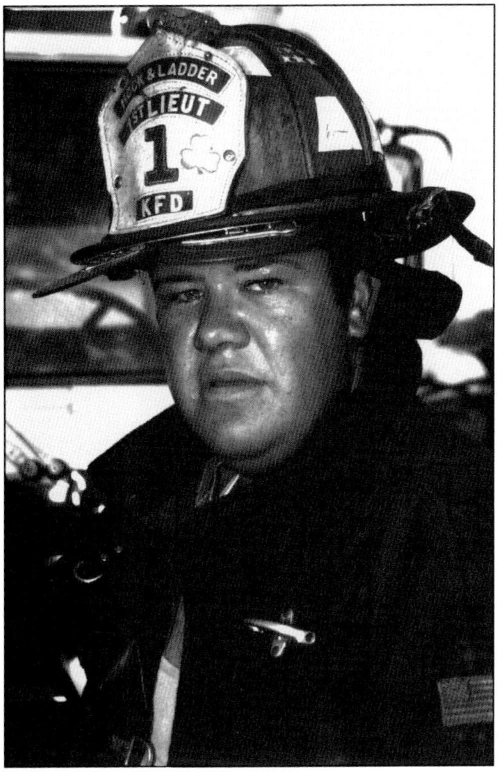

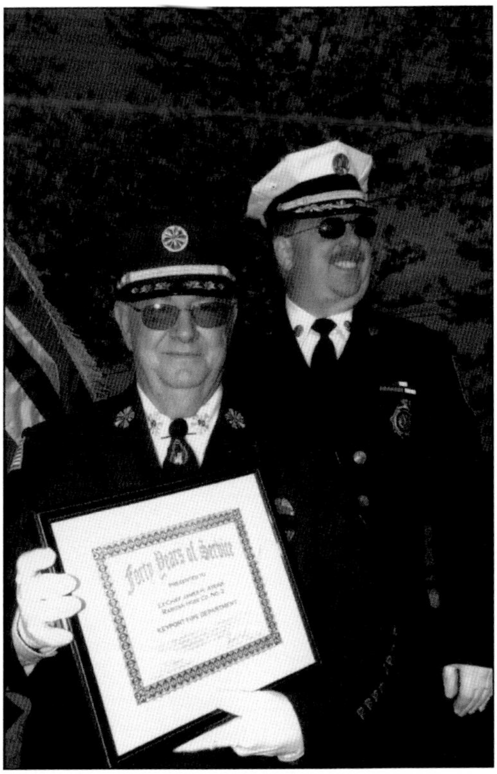

FORTY YEARS FOR JAMES ATKINS. James Atkins joined the KFD as an active member in 1963. He spent his first years at Eagle Hose Company, where he became an accomplished fire photographer and skilled firefighter. Atkins later joined Raritan Hose and served as chief of the KFD in 1990. In this photograph, he is joined by his longtime friend and chief of the KFD in 2007, William Larkin of the Eagle Hose Company. (Photograph by Pete Massas.)

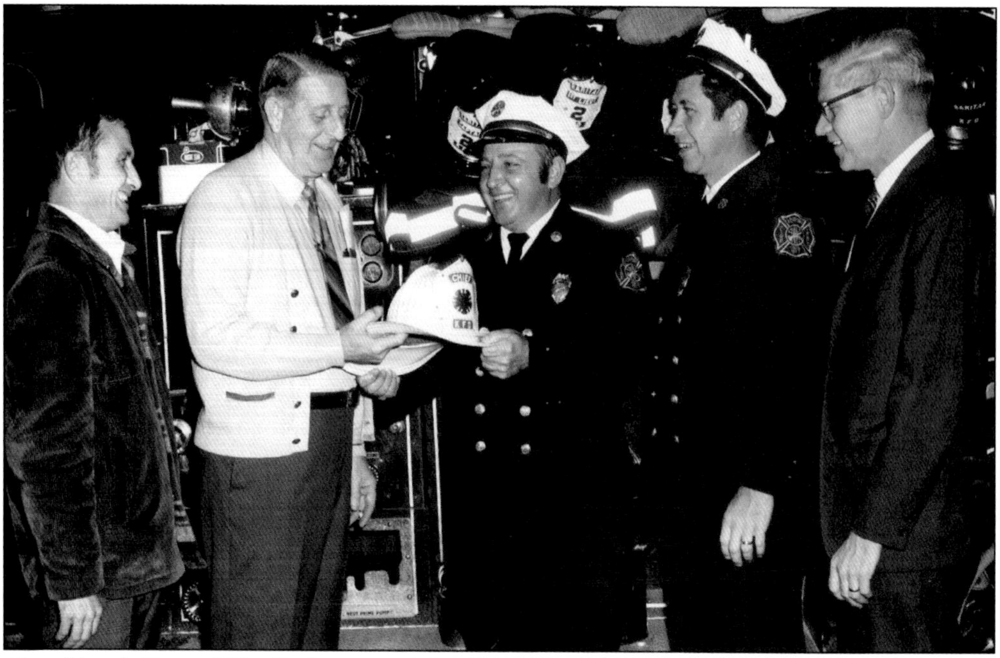

NEW YEAR'S DAY, 1971. Each year on New Year's Day, the Board of Chiefs takes their official positions for the following year. Here ex-chief Bob Rose hands off the chief's helmet to Harold Dorl while, from left to right, Assistant Chiefs Gene Regan and Ray Walling and firefighter Bob Redmond look on. (Photograph by James Atkins.)

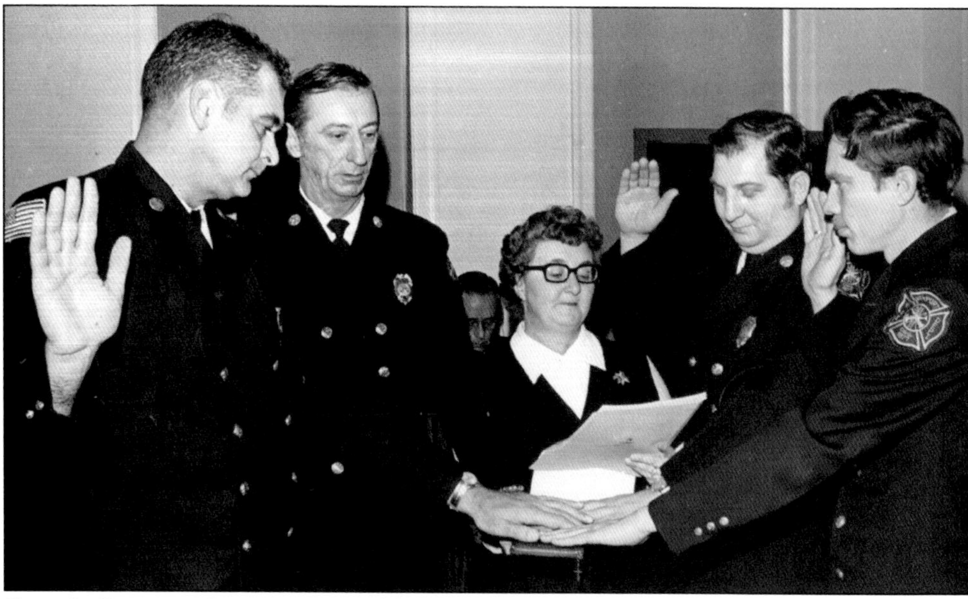

CHIEF'S OATH OF OFFICE, 1974. Although Keyport has an all-volunteer fire department, it is municipally controlled, operated, and funded. Each year, the Board of Chiefs must swear under oath to uphold the constitution of New Jersey and the laws of the Borough of Keyport. Seen here are, from left to right, Second Assistant Chief George Hewitt, First Assistant Chief William Nagle, borough clerk Barbara Tice, Chief John Kovacs, and Third Assistant Chief Harry M. Aumack II.

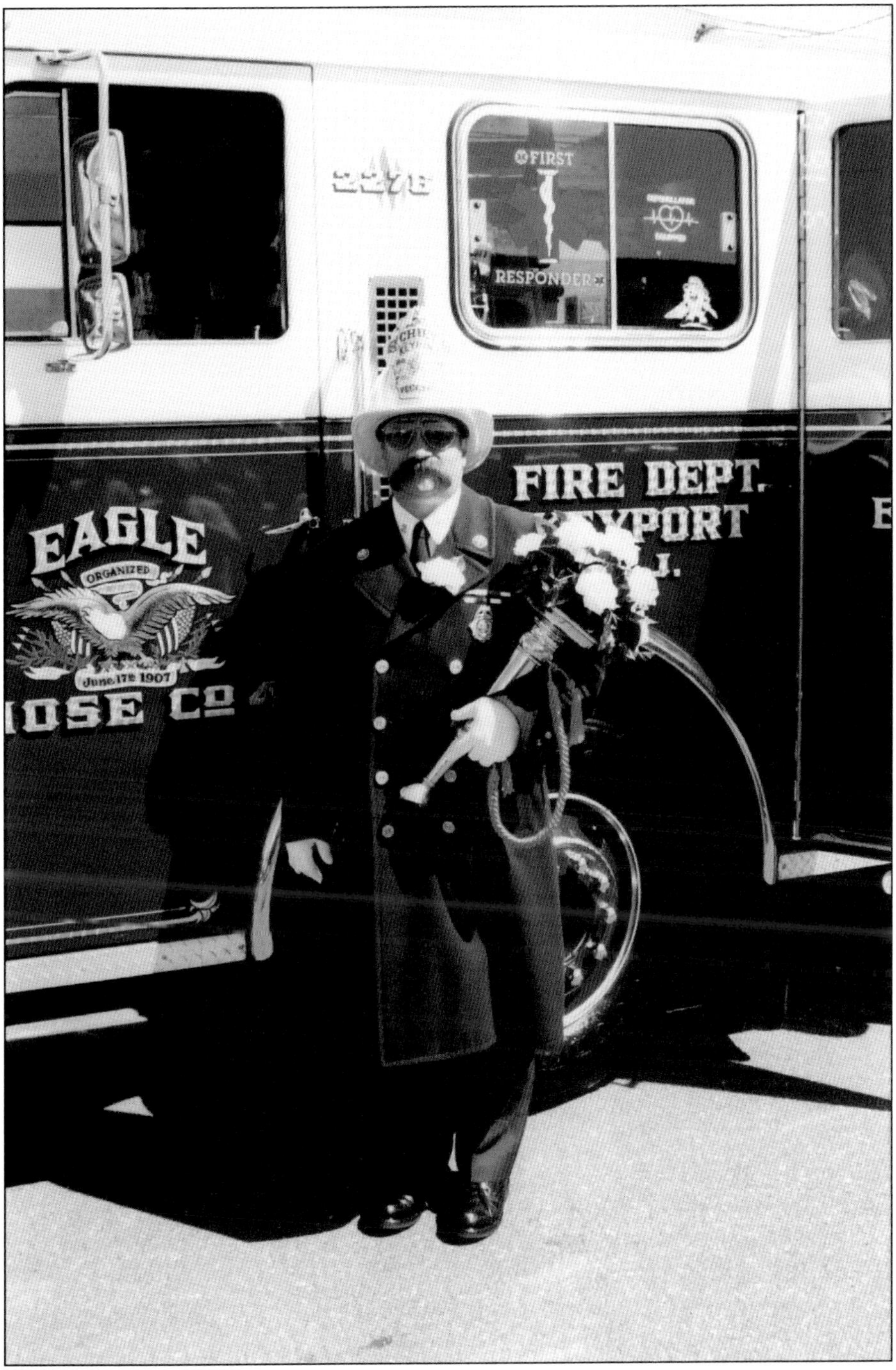

Chief Anthony P. Vecchio, 2000. Anthony P. Vecchio joined the Eagle Hose Company as an active member on December 1, 1976. He moved through the ranks and served as captain of Eagle Hose Company and later became the fire marshal for the Borough of Keyport. In 1997, he was elected to the Board of Chiefs under then chief William Jeandron. He is seen here dressed in his vintage 1890 chief's uniform, just before stepping off for an award-winning 2000 Fireman's Convention in Wildwood.

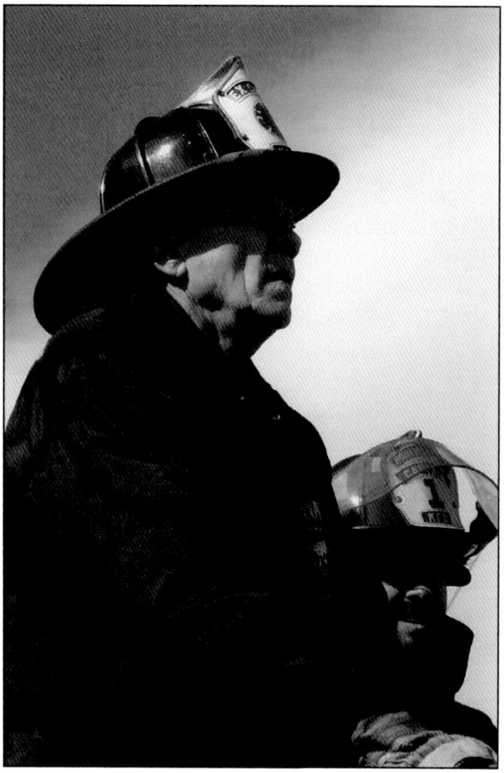

HARRY M. AUMACK II. Since the birth of the fire department in 1877, the Aumack family has been an integral part of it. This family became the first to have three successive generations be elected and serve as fire chief. Harry M. Aumack II watches over ladder operations with hook and ladder captain Michael Vinik.

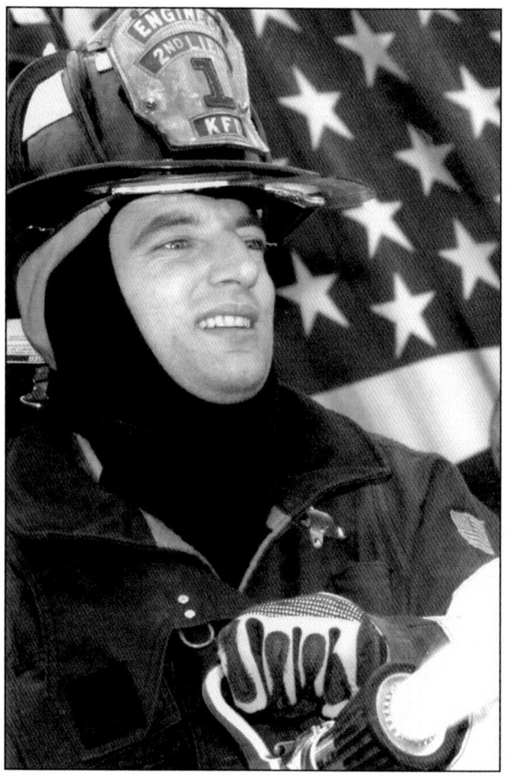

FIREFIGHTER GREG MADDOX. Keyport Engine Company member Greg Maddox was an accomplished sign painter and artist. A native of New Zealand, he served with the KFD for six years and during that time was responsible for painting the portrait of the September 11, 2001, New York City Fire Department flag raising on the side of the Keyport Engine Company Pierce Pumper.

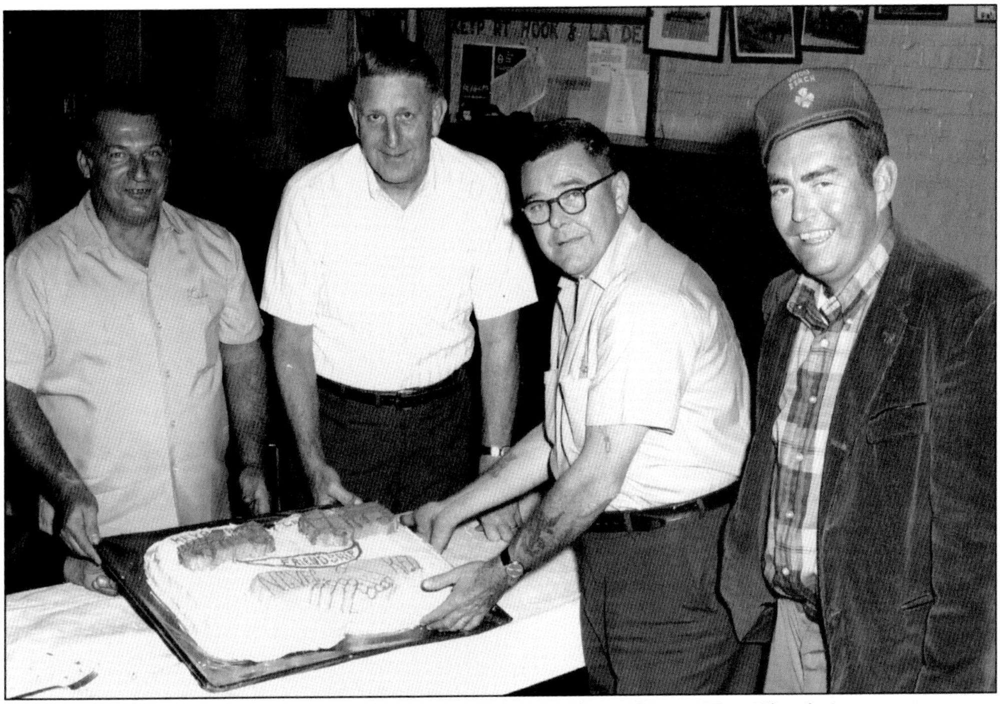

BAYSHORE CHIEFS, 1969. The first meetings of the Bayshore Active Fire Chiefs Association were held in the closing years of the 1960s. The one-year anniversary of this mutual aid organization was marked by a party at the old hook and ladder firehouse. Attending were Chief Merkel from North Centerville Fire Company, Chiefs Robert Rose and Howard Creed from Keyport, and Chief Jack Seber from Union Beach.

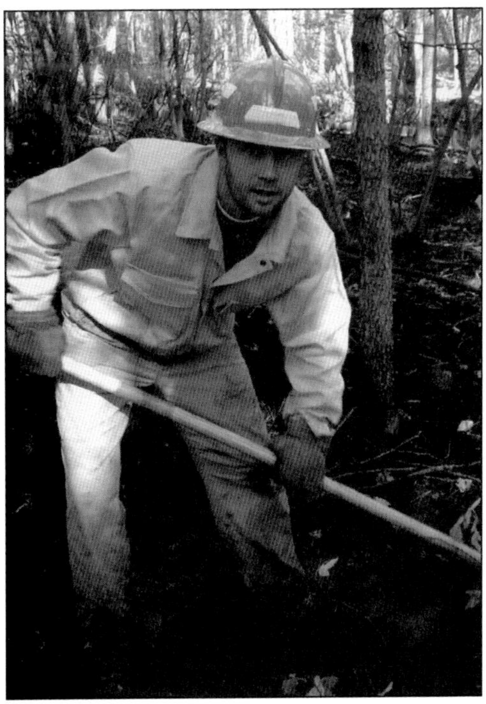

FIREFIGHTER DANIEL REGAN. Daniel Regan is shown here hard at work during a mutual aid forest fire in 2008 in Old Bridge Township. A 15-year member of the department and the Lincoln Hose Company, Regan was awarded several local and county valor awards for rescuing an elderly woman in a Second Street house fire.

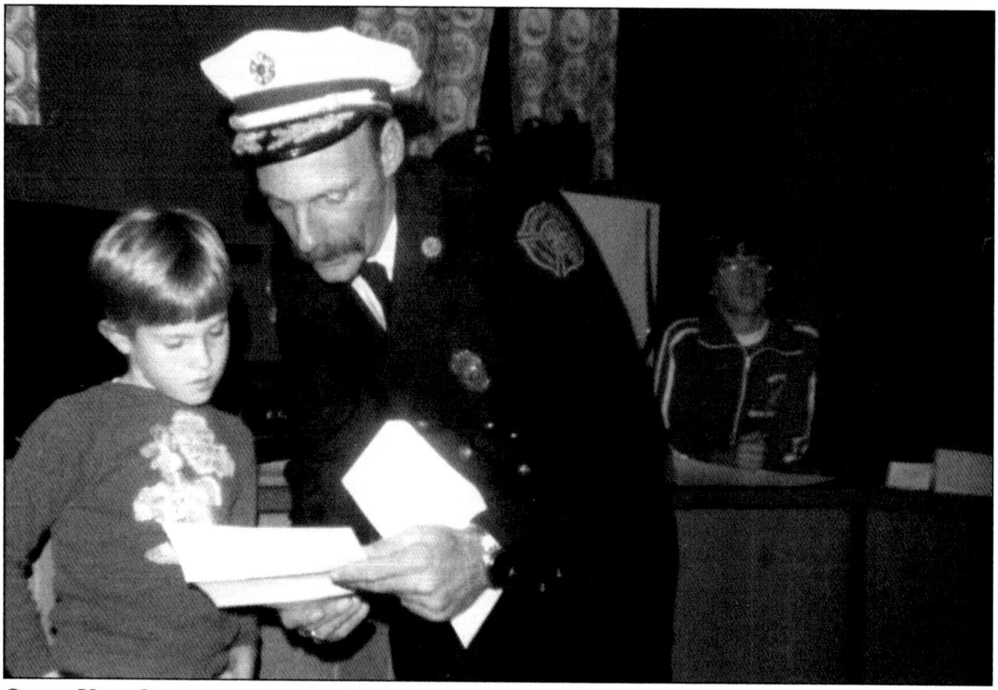

CHIEF KEN GOBLE, 1981. Chief Ken Goble joined the Board of Wardens in 1968 and soon served as captain at the Waverly Street firehouse. In 1978, he was elected to the position of third assistant chief and served as chief of the department in 1981. He is seen here congratulating a winner of the 1981 school fire prevention poster contest.

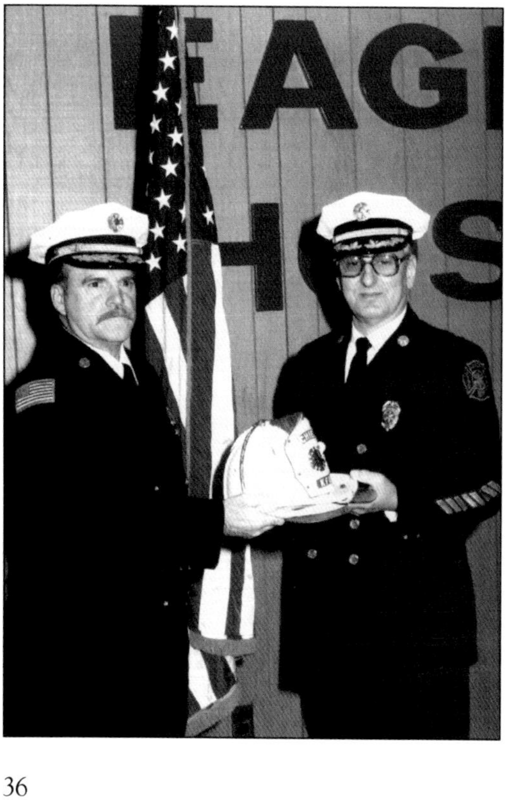

PASSING OF CHIEF'S HELMET, 1986. Each January 1, it is tradition for the outgoing fire chief to hand over the reins of the department. Here Robert "Pop" Poling hands off the chief's helmet to Julius (Jules) Rosato at Eagle Hose Company. Each fire chief serves a one-year term at each ascending rank from third assistant, to second assistant, then to first assistant, and finally to chief of the department for a total of four years on the Board of Chiefs.

FIREFIGHTER JOHN MCCLEASTER. John McCleaster joined the Liberty Hose Company in 1980. He has served as the department mechanic and as captain of Liberty Hose Company. As of 2009, he was elected to serve as the chief of the KFD in 2012.

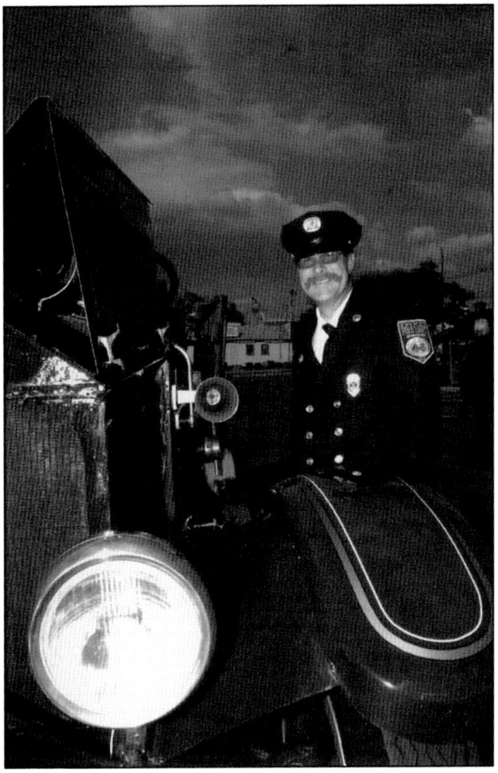

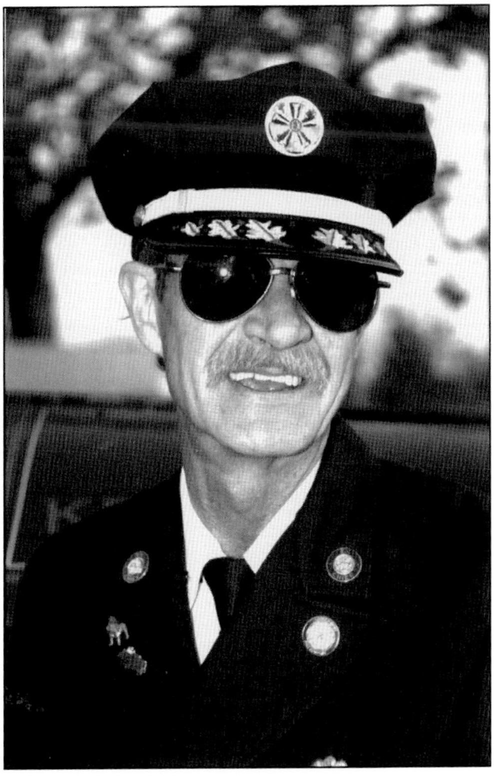

EX-CHIEF JAMES ENGLISH. Lincoln Hose Company member James English joined the KFD in 1975. He served as captain of Lincoln Hose Company and for over 30 years has been one of the most active Keyport firefighters. In 1999, English proudly served as chief of the KFD.

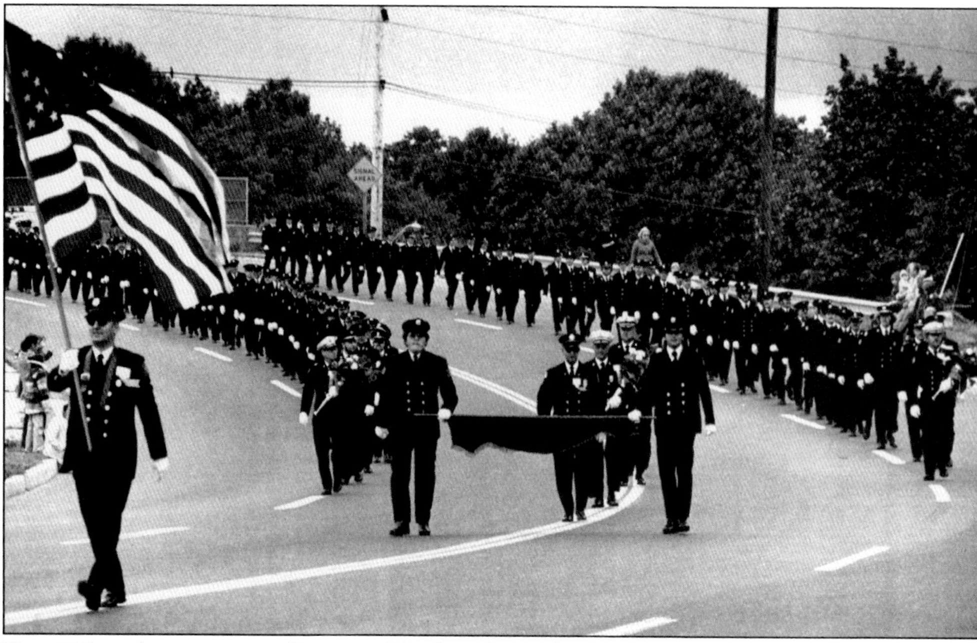

THE 100TH ANNIVERSARY PARADE. Heading down the West Fourth Street overpass is the KFD marching in its 100th anniversary parade. The KFD has celebrated major anniversaries in 1927, 1952, 1967, 1977, and 2002. The most recent parade drew over 200 fire apparatus and nearly 5,000 persons who came to help the KFD with its 125th birthday.

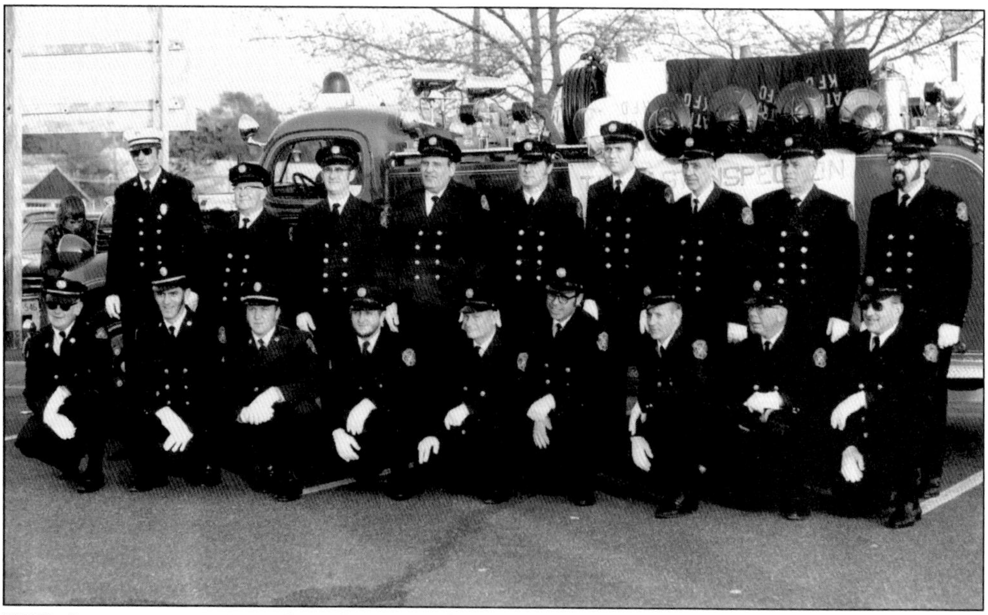

KFD INSPECTION, 1972. Seen here are, from left to right, (kneeling) Ed Keeler, Bob Ferry, Louis Latorette, Eddie Bennett, Ted Barber, Vito Esposito, Frank Flynn, Leroy Stultz, and Richie Poling; (standing) Assistant Chief William Nagle, Frank Van Pelt, Earl Hallem, Matt Danza, William Siegle, Paul Rudolph, Don Vaughn, Stan Ferry, and Harold Poling. This would be the last inspection for the Keyport Fire Patrol's 1946 Dodge Tasc, which was soon to be replaced by a new Hahn pumper.

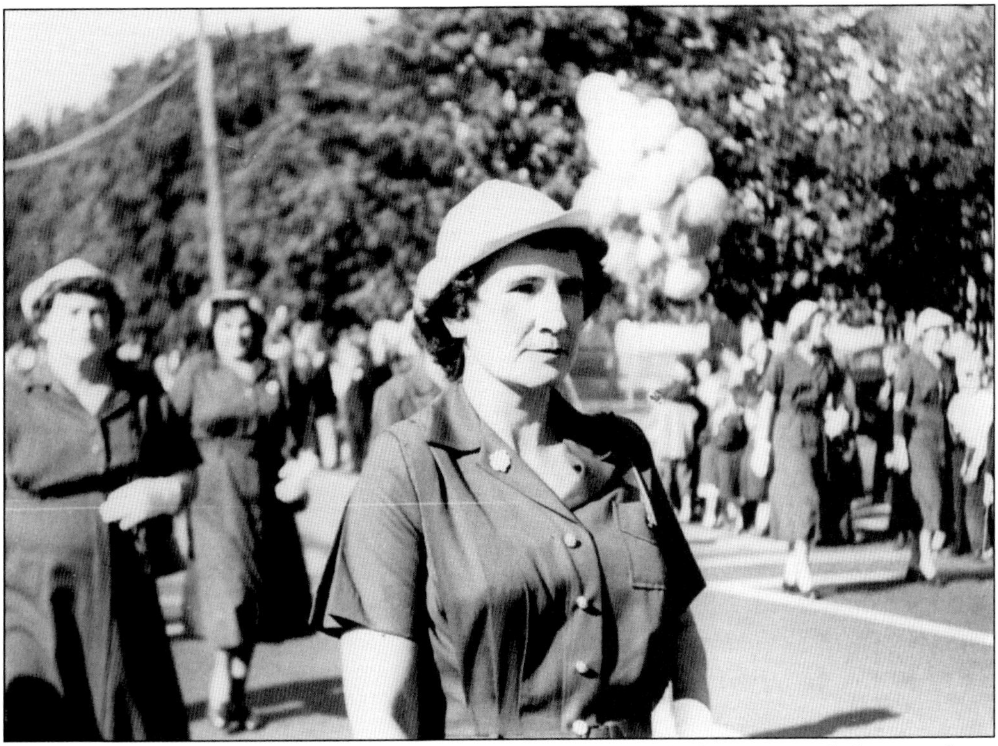

Therese Rose, 1949. Here Therese (Tess) Rose proudly leads the Keyport ladies marching unit in a 1949 parade. Her husband J. Robert Rose served with the Keyport Hook and Ladder Company for over 40 years and was chief of the department in 1970. Tess spent over 50 years serving as an auxiliary member of the Keyport Hook and Ladder Company.

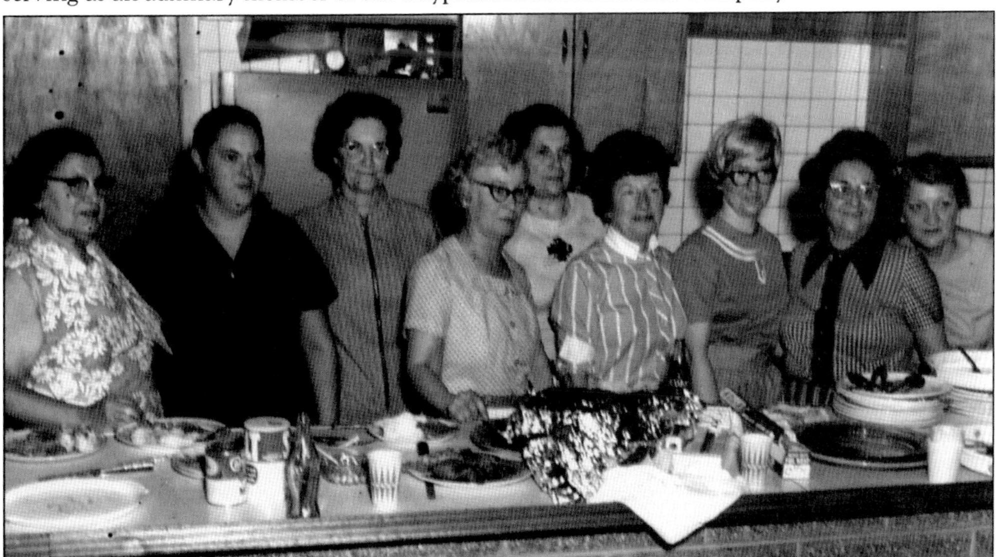

Hook and Ladder Auxiliary, 1968. Keyport's first ladies auxiliary was formed in 1921 by concerned wives of the Keyport Hook and Ladder Company. Seen here are, from left to right, Angie Diodato, Lenore Dente, Fay Bennett, Hattie Walling, Tess Rose, Marge Boyce, Pat Davison, Ann Kitzman, and Elsie Roberts.

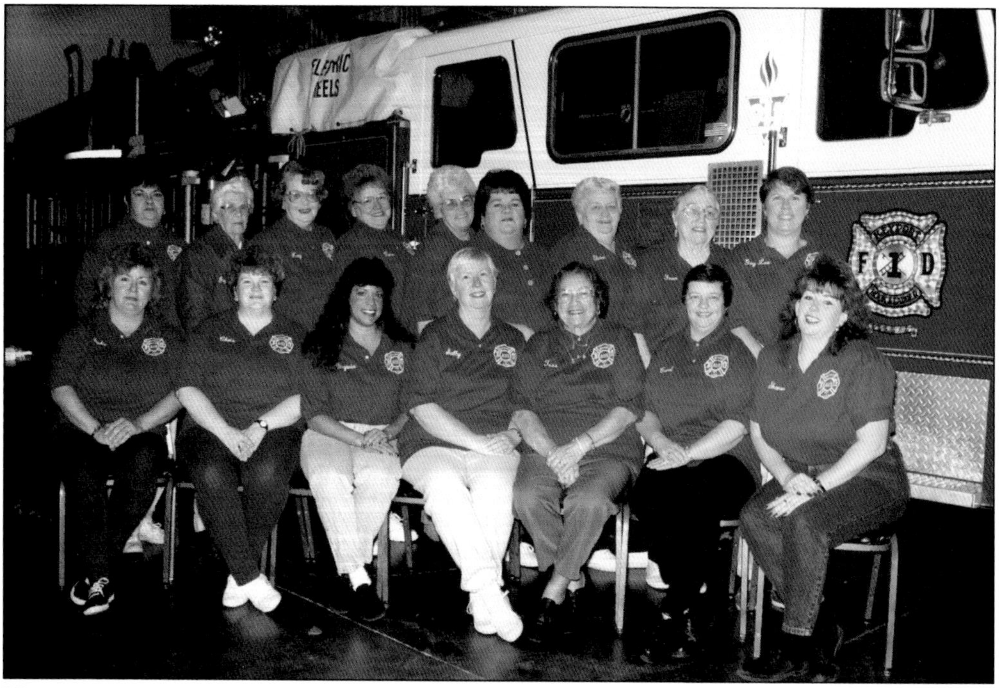

HOOK AND LADDER AUXILIARY, 1996. Seen here are, from left to right, (first row) unidentified, Christine Aumack, Virginia Maldonado, Sally Kitzman, Tess Rose, Carol Mills, and Sharon Graham; (second row) Lenore Enter, Millie Aumack, Kay Sorber, Rose Borghi, Fay Phillip, Nicole Caddle, Elaine Aumack, Ran Ross, and Gaylee Benedict.

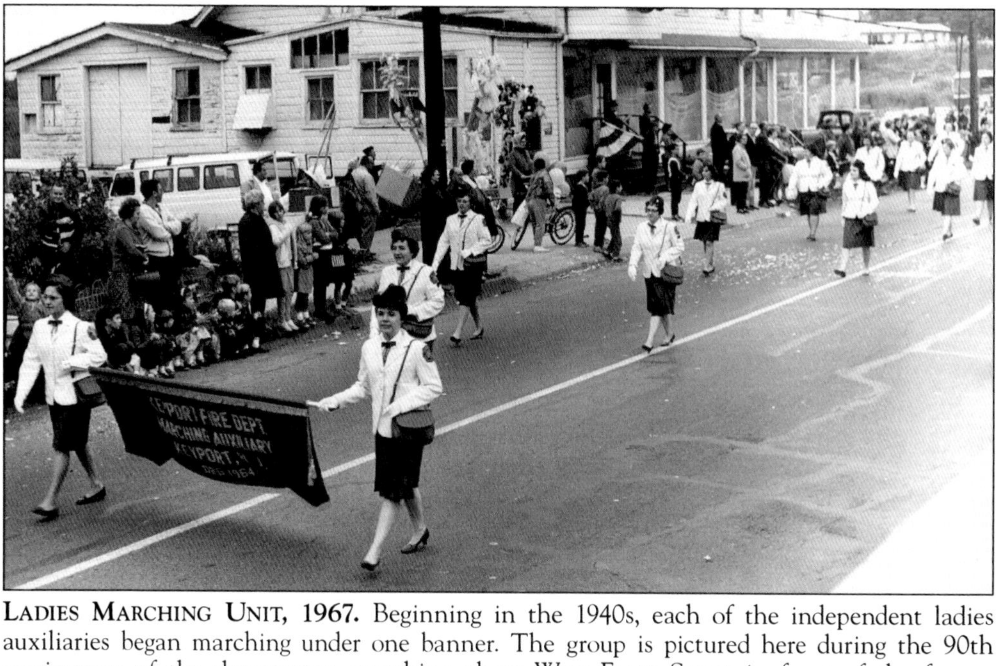

LADIES MARCHING UNIT, 1967. Beginning in the 1940s, each of the independent ladies auxiliaries began marching under one banner. The group is pictured here during the 90th anniversary of the department, marching along West Front Street in front of the former Oyster House. In just a few years, the historic Oyster House would be consumed by a midnight general-alarm fire at pull box No. 41.

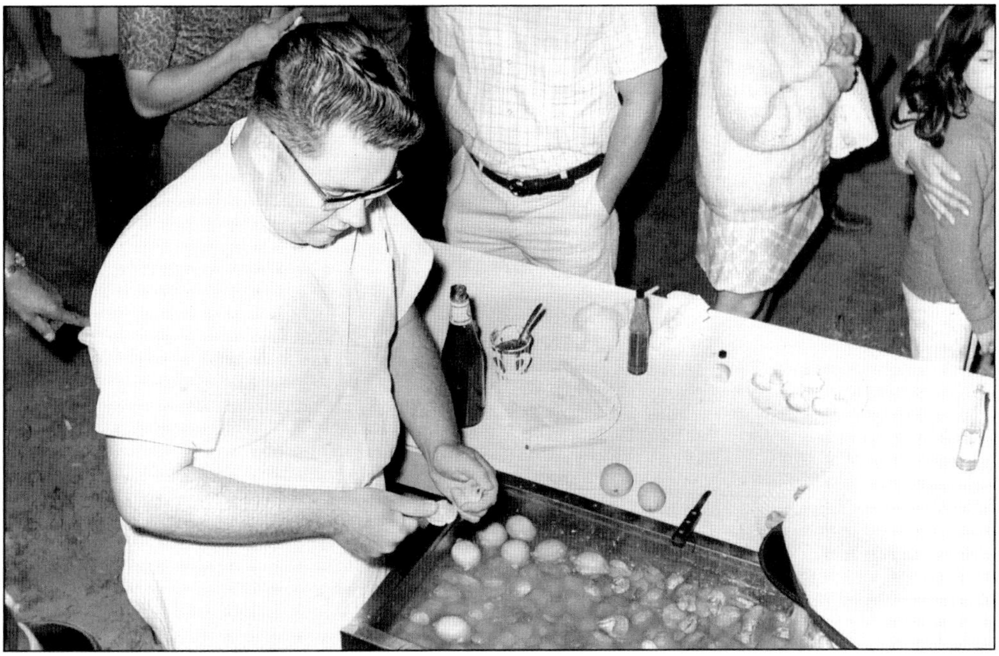

FIREMEN'S FAIR, 1969. Each year, the fair takes an entire day to set up, runs from Monday through Saturday, and then breaks down on Sunday. Countless hours are needed by all firefighters, auxiliary members, and junior firefighters to make the event a success. Here Bob Wilson of Raritan Hose Company shucks clams on the half shell. Add some hot sauce, and boy is that a treat. (Photograph by David Dorl; courtesy Raritan Hose Company.)

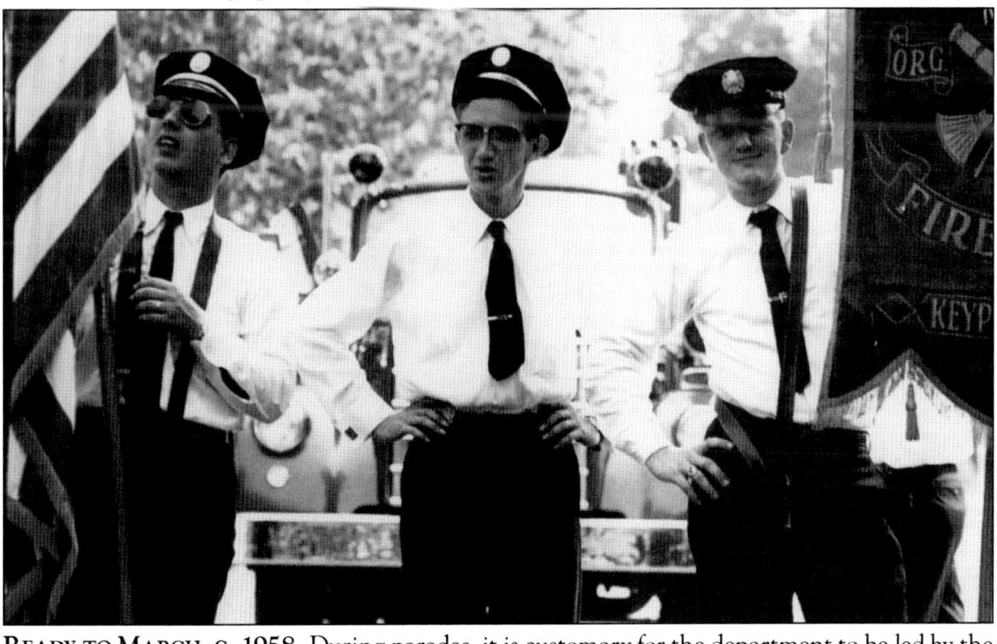

READY TO MARCH, C. 1958. During parades, it is customary for the department to be led by the banner and color guard. For many years, the single-staff banner pictured here was carried until being replaced by the current two-person horizontal banner. From left to right are Robert Holtz, longtime department secretary and historian; Bill Phillip; and Ken Young.

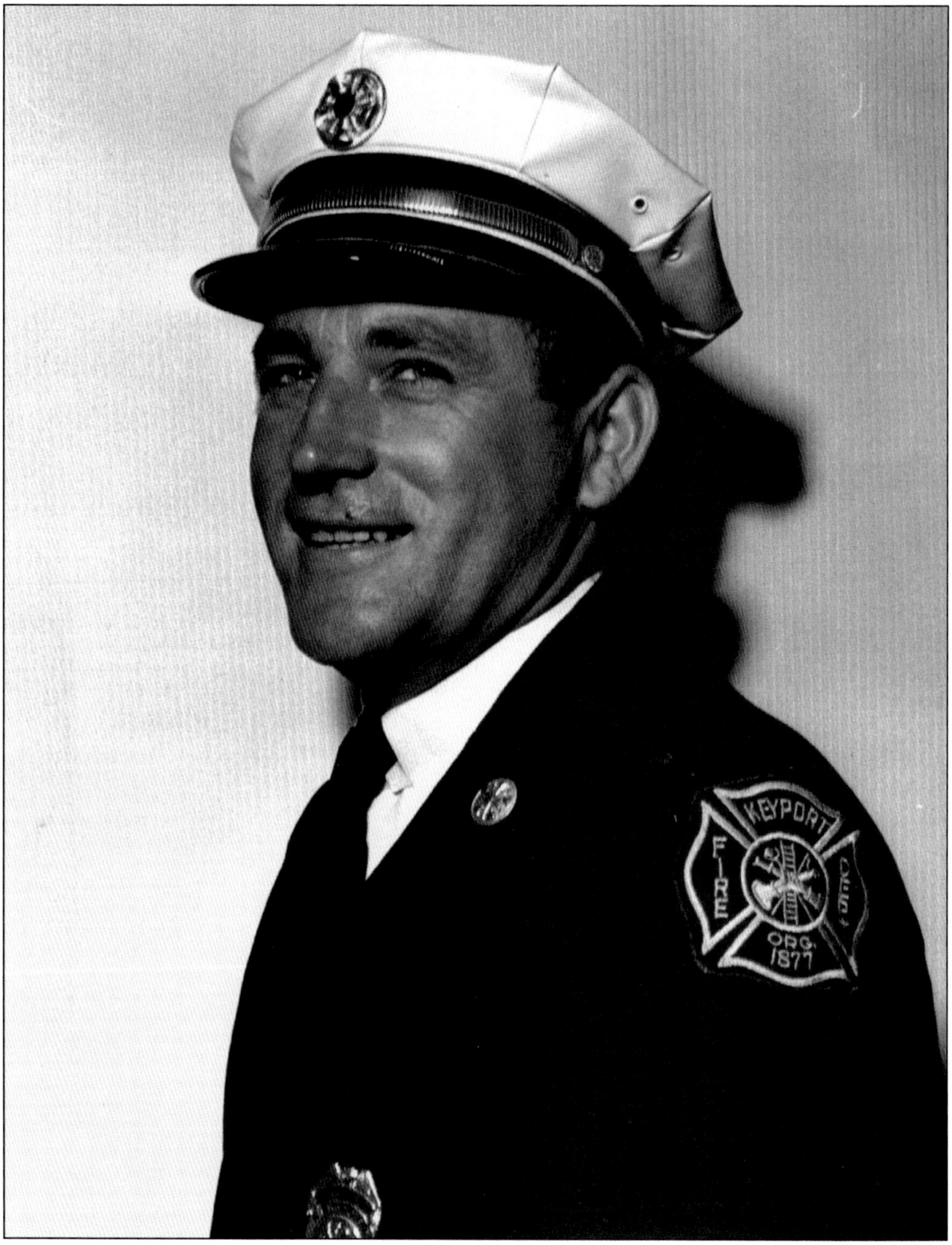

CHIEF FRANK "JUNIOR" VAN PELT, 1967. Frank Van Pelt joined the Lincoln Hose Company in 1951 while still attending Keyport High School. Known as "Junior" throughout the department, he was instrumental in forming the Keyport Fireman's Fair in 1967, as well as playing key roles in the department's 90th and 100th anniversaries and the 1982 Exempt Parade, and was chairman of the 125th department anniversary at the time of his passing.

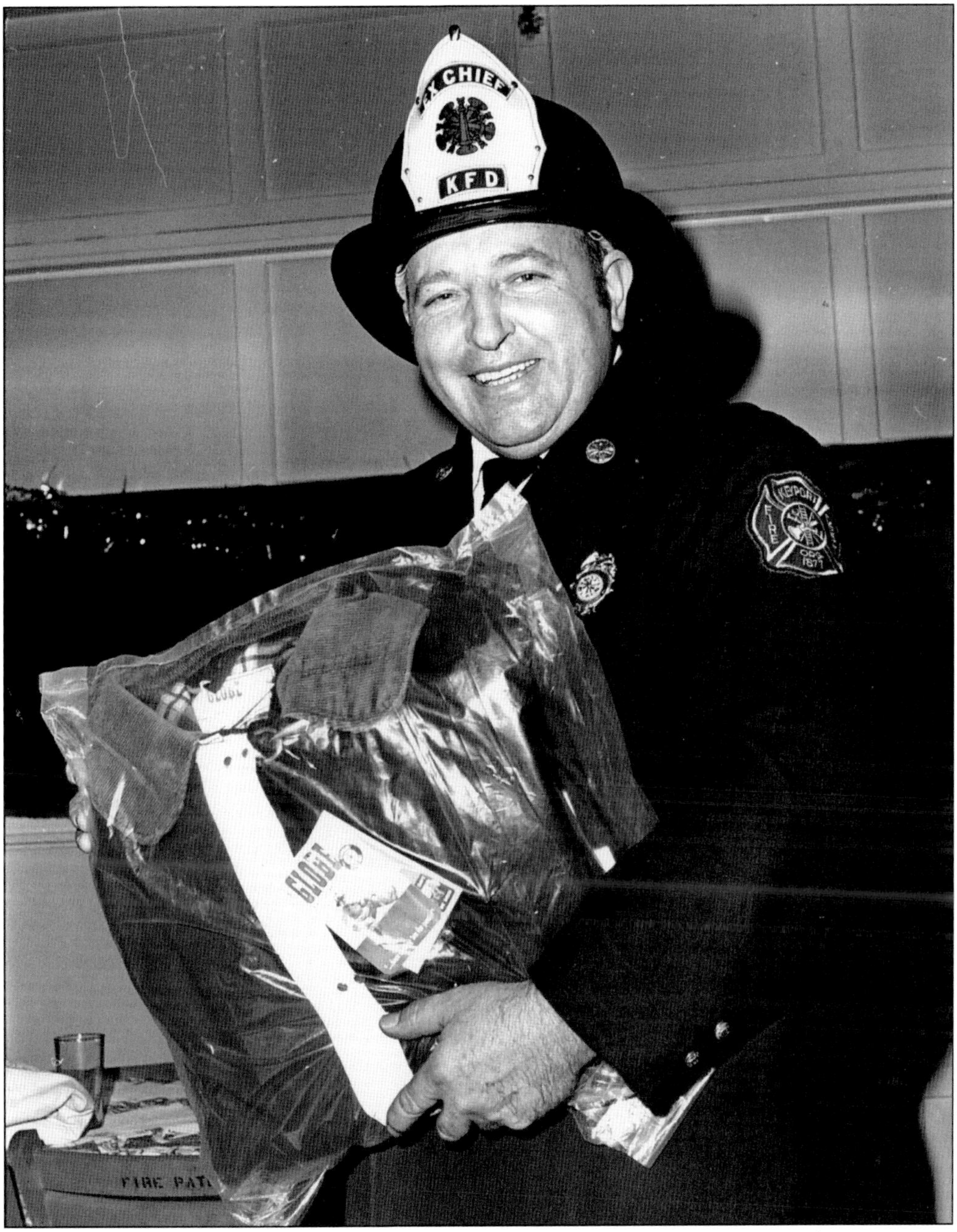

EX-CHIEF DORL, 1972. The Raritan Hose Company accepted Harold E. Dorl as an active member on December 10, 1941. Firefighter Dorl served in all positions of the KFD, becoming chief of the department in 1971. He is accepting his ex-chief coat and helmet, which traditionally took place on New Year's Day for many years. (Courtesy Raritan Hose Company.)

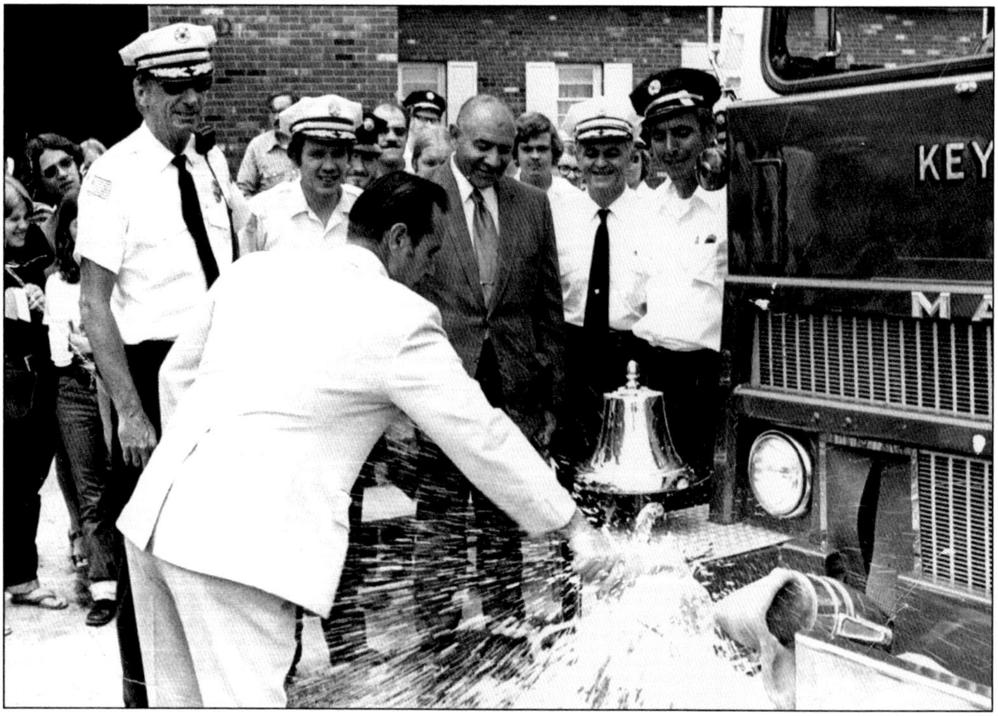

ENGINE DEDICATION, 1974. With a crash, Mayor William Ralph dedicates the new Mack truck of Raritan Hose Company. After the christening by champagne, the engine will be blessed by the chaplain, praying for many years of good luck and safety. Looking on are, from left to right, Fire Chief Bill Nagle, Assistant Chief Harry M. Aumack II, councilman Don Miller, Assistant Chief Leo McCann, and ex-chief Bill Newman. (Courtesy Raritan Hose Company.)

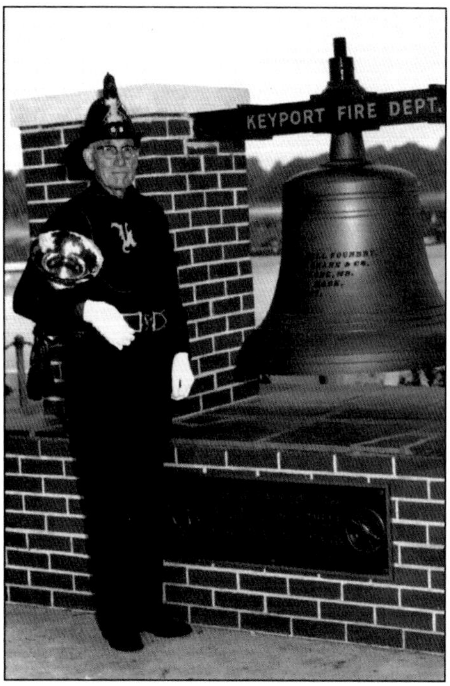

EX-CHIEF GEORGE BIRCH, 1962. George Birch joined the ranks of the Raritan Hose Company in December 1907. He is shown here during the dedication of the Fireman's Monument in 1962 wearing the uniform of his father, George Sr., who was a member from 1878 until 1899. George watched the 1963 inspection parade from his First Street porch and passed away on Christmas day 1963 after 56 years of volunteer service.

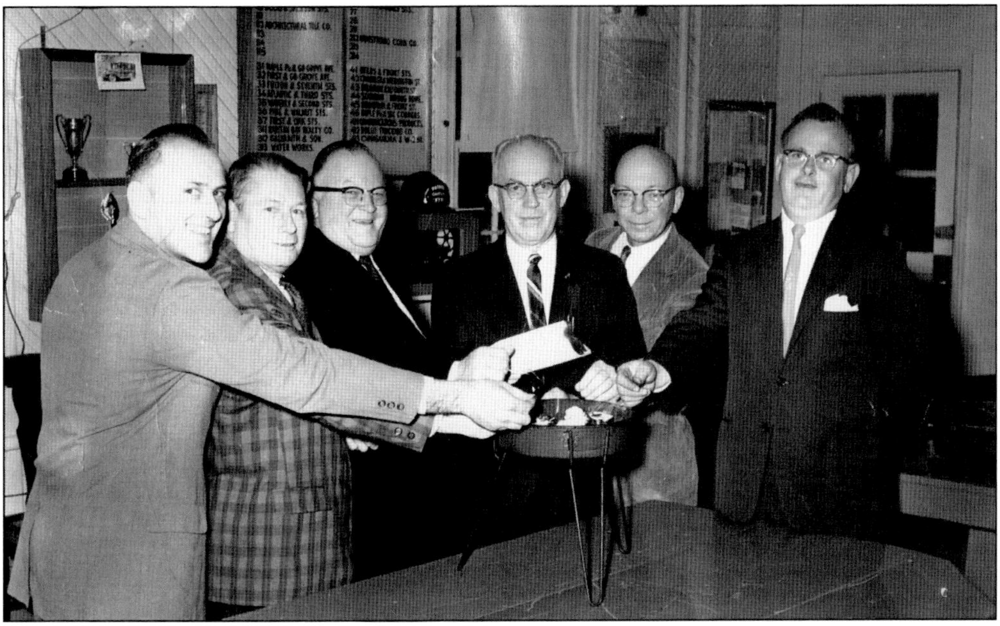

PATROL MORTGAGE BURNING. In 1952, the Keyport Fire Patrol purchased the former firehouse of Lincoln Hose Company on Waverly Street. Within 13 short years, the mortgage note was paid off and burned in the firehouse in 1965 during a traditional ceremony. Gathered that day are, from left to right, Richie Poling, Ted Barber, Frank Van Pelt, Everett Poling, Leroy Stultz, and Ed Keeler.

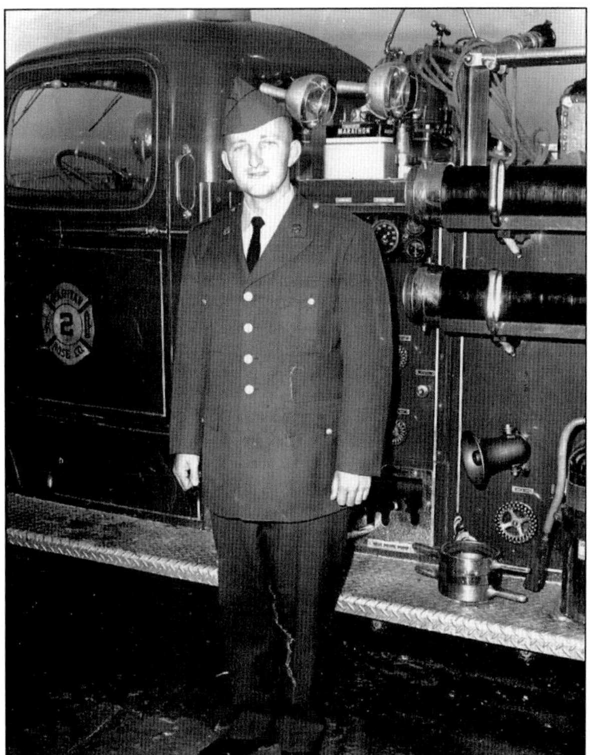

FIREFIGHTER JOHN "TED" LARKIN. John "Ted" Larkin first joined the Eagle Hose Company in 1964. Larkin is seen here just before he shipped off to Vietnam with the U.S. Army on November 14, 1968. Larkin was also one of two official chief's aides of the KFD: he in 1967 for Frank Van Pelt and the other, Randolph Stout, who served as Chief Bert Morris's aide in 1991.

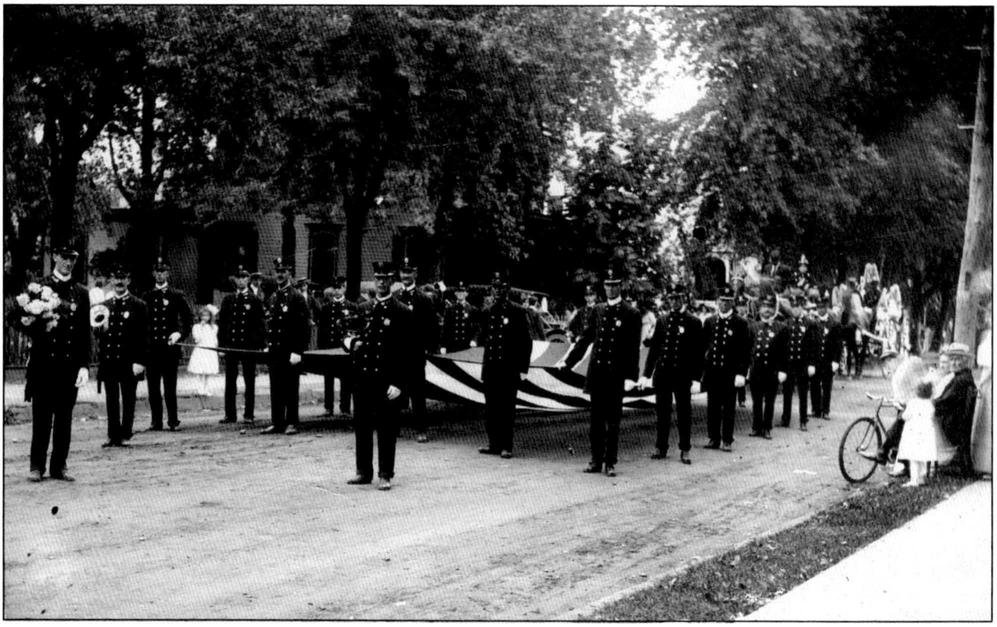

THE COLORS ON PARADE, 1912. For several years in the first part of the 20th century, Keyport firefighters carried this large flag in many local parades. During the 125th anniversary of the department in 2002, an identical flag was purchased and is now carried again in special events. (Courtesy Jim McTernan.)

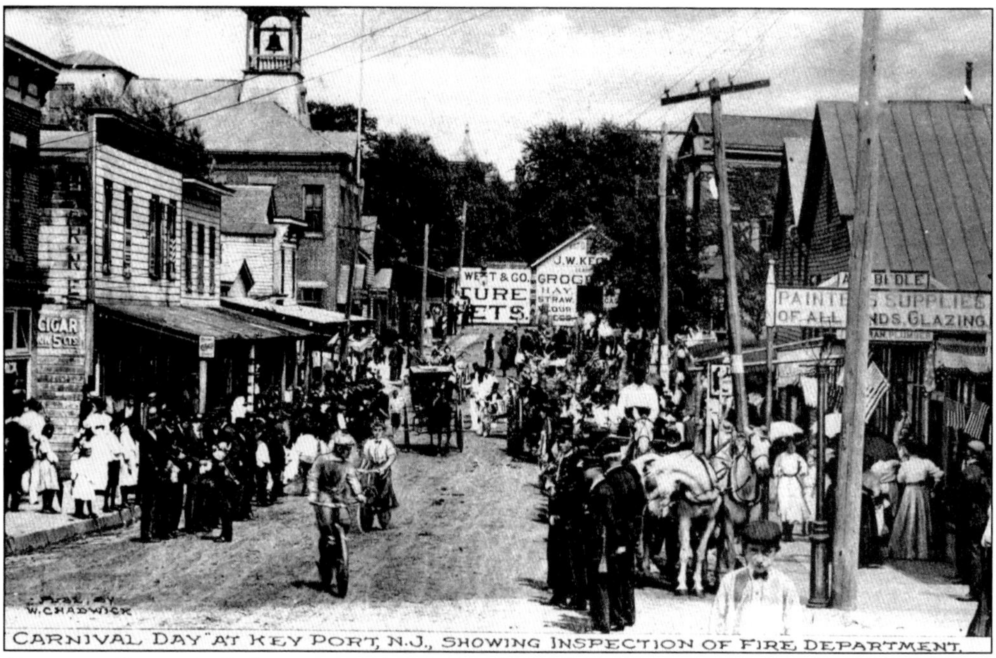

CARNIVAL DAY, 1905. Approximately 100 members of the department have turned out this day for the annual inspection on East Front Street. Inspection day has been an important day each year since 1891, when it was first held by Chief George W. Walling to determine the preparedness and fitness of each of the individual companies.

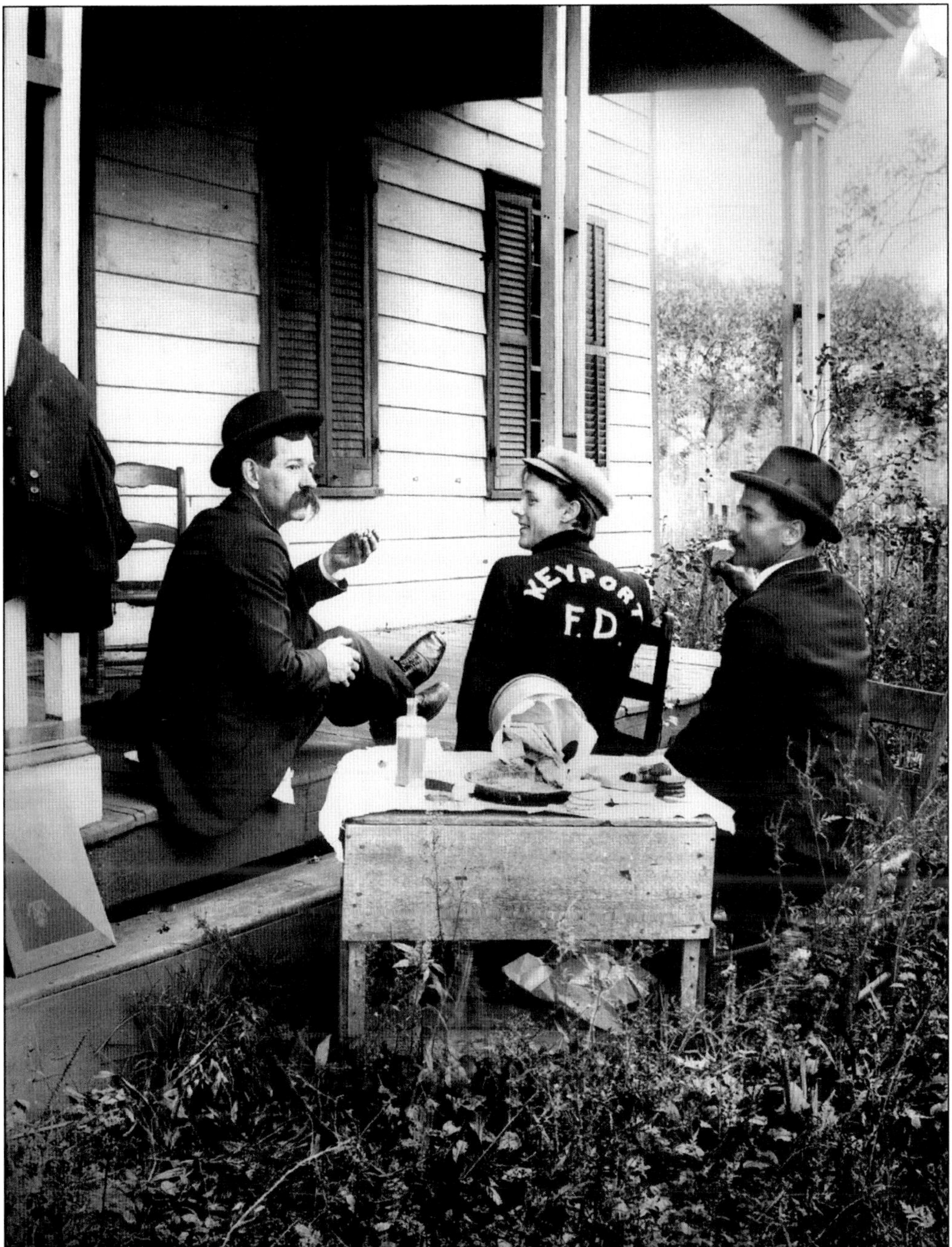

KEYPORT FIREMAN, C. 1890. Here a group of men, most likely all Keyport firefighters, are posing for the camera while eating cookies. The significance of this photograph lies in the embroidered sweater of the young man in the center. It was not known until this photograph surfaced that such "off duty" clothing was worn during that time. Since the late 1940s, all Keyport firefighters have had casual jackets and shirts boldly showing apparatus and company names that are proudly worn as an outward expression of members' pride and honor.

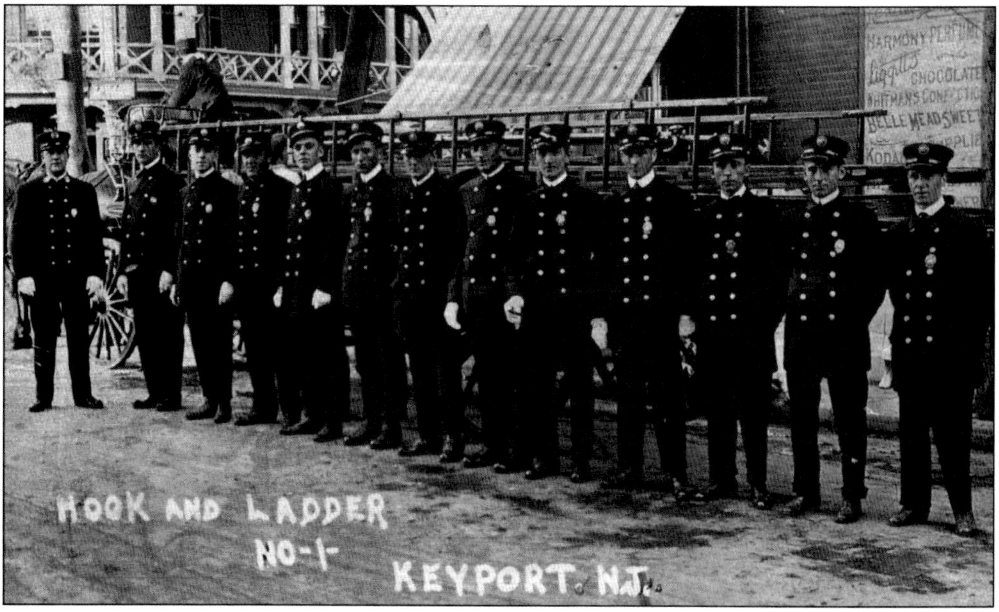

HOOK AND LADDER COMPANY NO. 1, 1915. Nearly all of Keyport's "truckies" have turned out for this annual inspection and parade. In firefighting terms, companies that operate ladder trucks or aerials are known as truck companies, and the members are affectionately nicknamed truckies. Technically, when referring to a fire truck, it would be one with some type of ladder or aerial device.

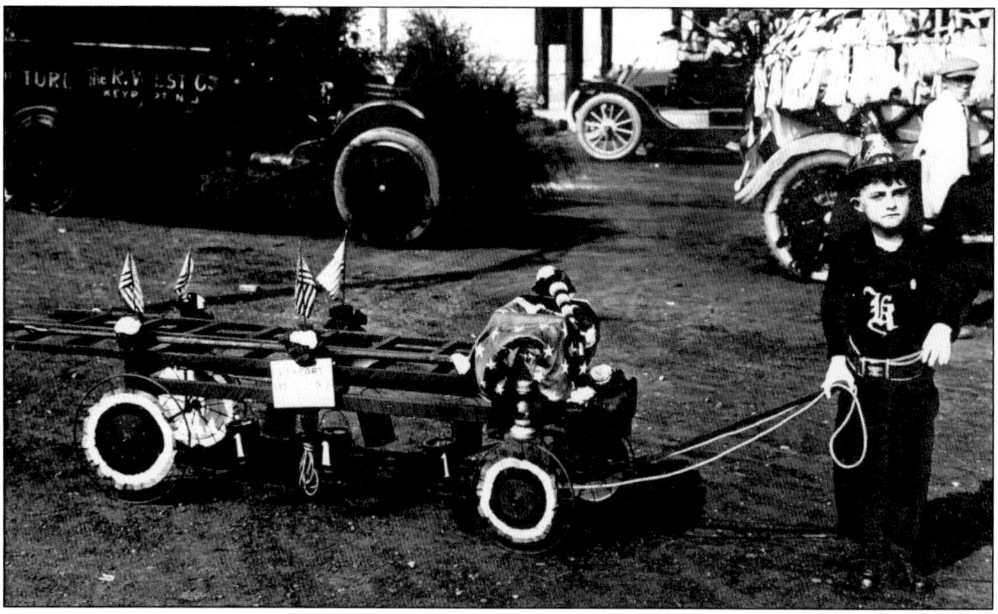

FUTURE HOOK AND LADDER MEMBER, 1912. For so many young men, running with the engines was a pastime that would eventually lead to volunteer membership. Here young Jack Heyer is ready to take the ropes of his miniature hook and ladder wagon. Complete with all the tools and equipment of its big counterpart, this little Seagrave is ready for the streets. Jack is wearing the uniform of his grandfather John H. Heyer, who was a member of the company from October 31, 1877, until January 6, 1910.

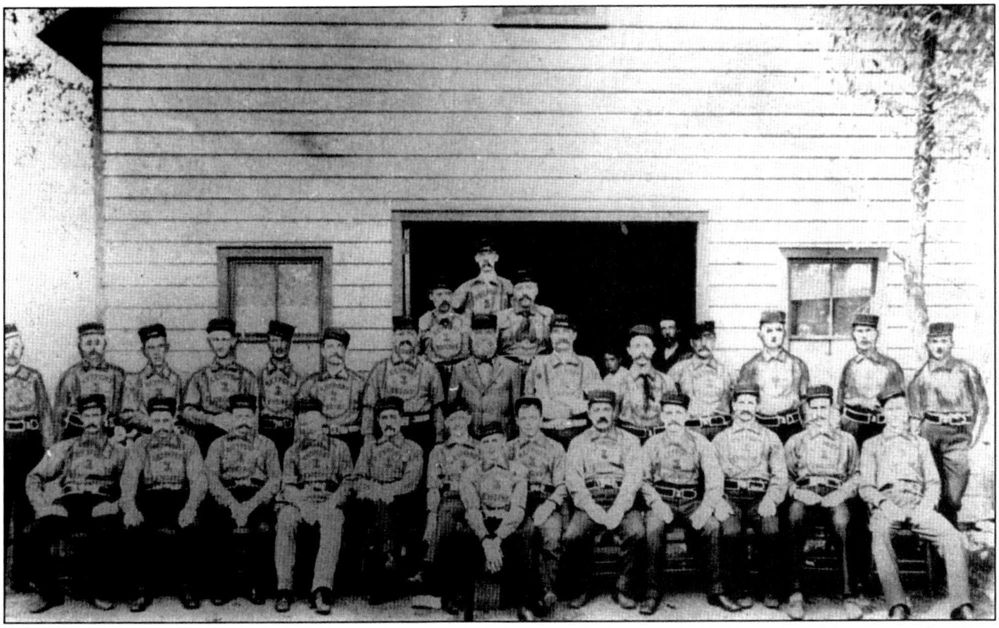

ENGINE COMPANY NO. 1, 1891. The whole clan has turned out for what would become the first portrait of Engine Company No. 1 in 1891. In the center is Chief George Washington Walling, who was born in Keyport and later served as police superintendent for the City of New York. He retired to his hometown and was appointed fire chief due to his overwhelming respect and public safety knowledge. He served for just a few short months when he contracted pneumonia and died at home.

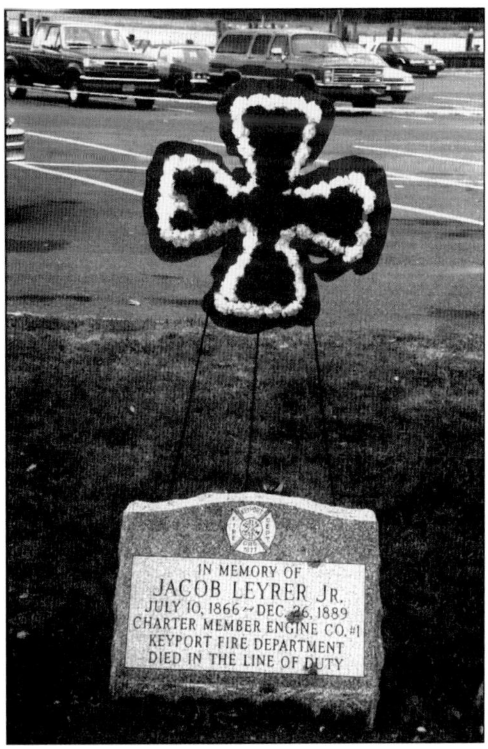

FIREFIGHTER LEYRER TABLET. In 1889, Engine Company No. 1 member Jacob Leyrer was killed when his own family's bakery caught fire on Thanksgiving Eve 1889. The tragic event took the life of Jacob's father and young Jacob, who was badly burned while attempting to save his father. After nearly a month of clinging to life at the Pavilion Hotel he died on December 26, 1889. (Courtesy Tim Regan.)

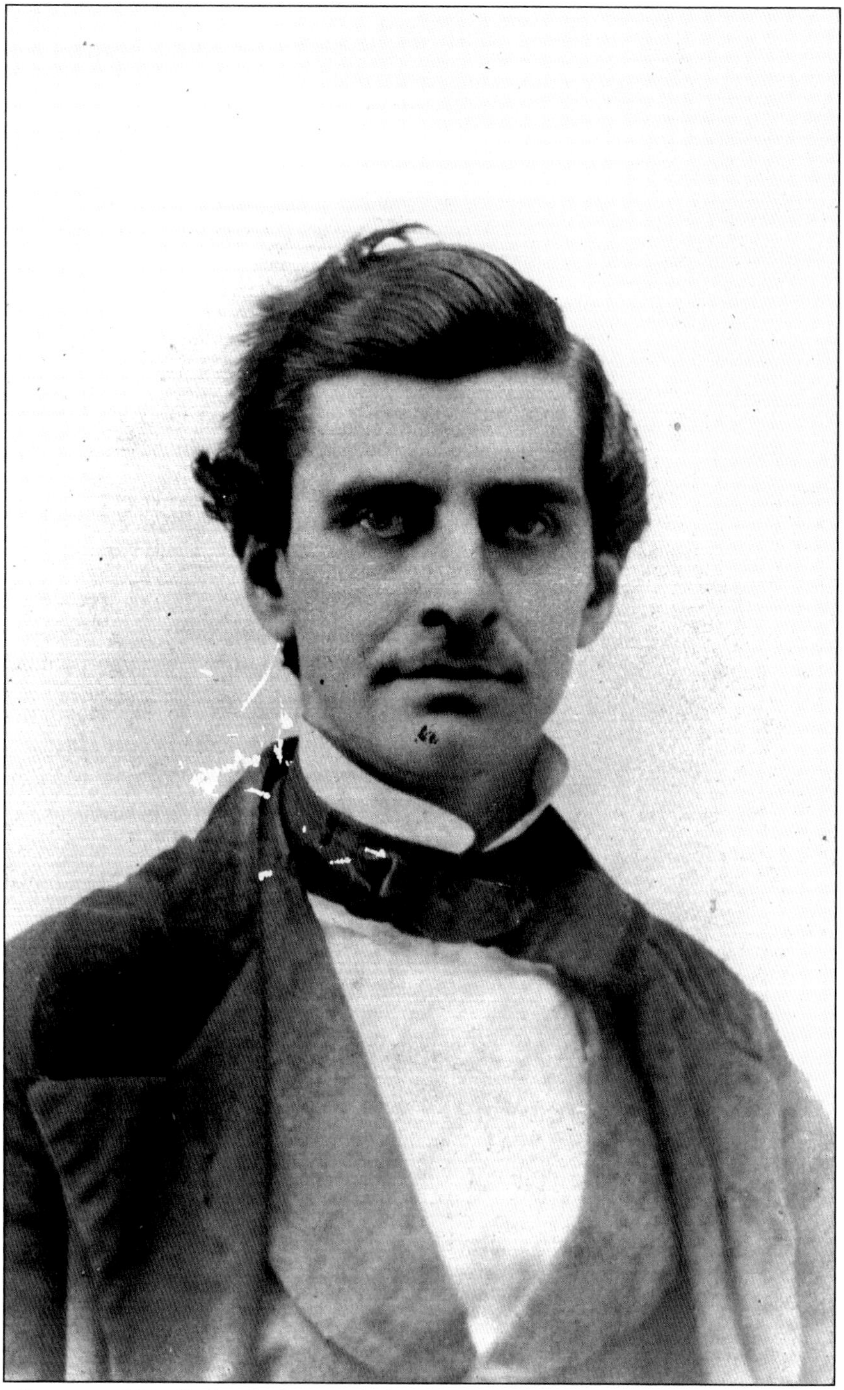

MAYOR OGDEN, 1870. Rufus Ogden settled in Keyport in 1850, when he temporarily set up a harness shop on West Front Street. Over the next 60 years, he served as the first mayor, businessman, charter member of the fire department, builder of the first town hall, and firehouse and town historian. Although his stay in town was "temporary" his improvements to the community were long lasting.

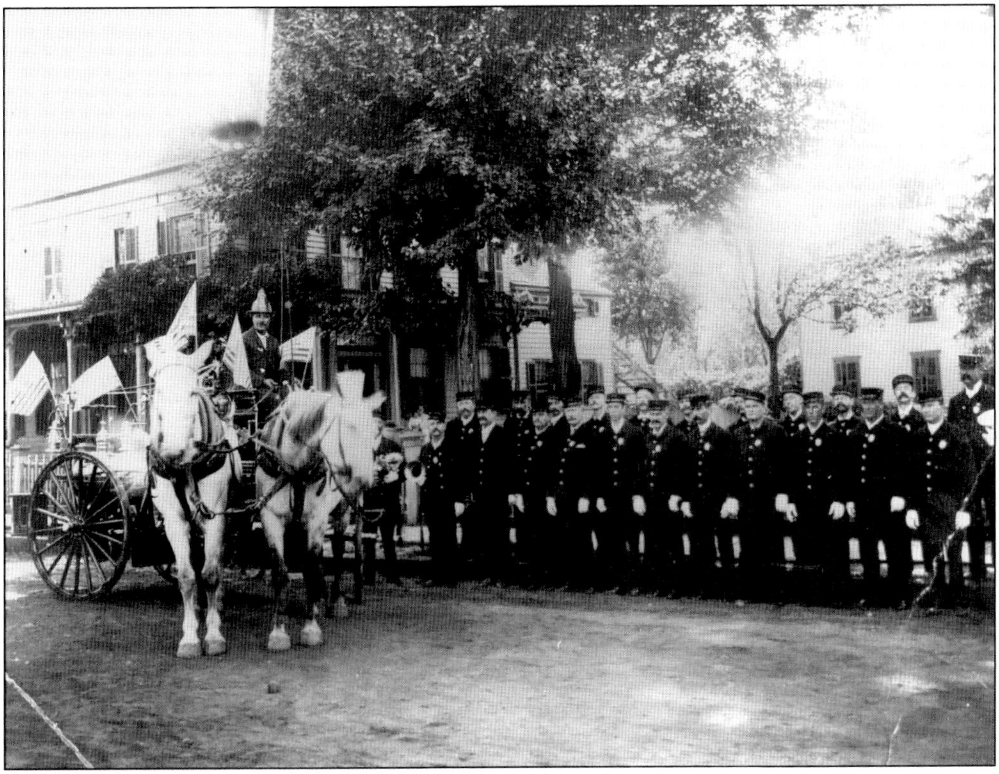

Keyport Engine Company, c. 1906. Sporting their class A uniforms, the Keyport Engine Company boys have gathered on Main Street near West Front Street for a formal portrait. Many of these men were among those at the Leyrer bakery fire, which claimed the life of fellow Keyport Engine Company member Jacob Leyrer Jr.

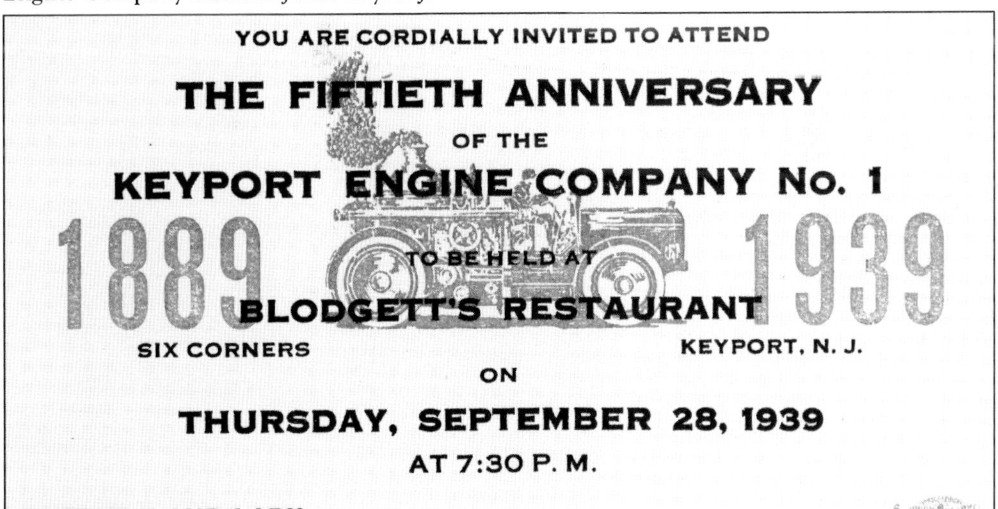

YOU ARE CORDIALLY INVITED TO ATTEND

THE FIFTIETH ANNIVERSARY

OF THE

KEYPORT ENGINE COMPANY No. 1

1889 1939

TO BE HELD AT

BLODGETT'S RESTAURANT

SIX CORNERS KEYPORT, N. J.

ON

THURSDAY, SEPTEMBER 28, 1939

AT 7:30 P. M.

YOURSELF AND LADY

Invitation Card, 1939. One did not dare to be late to this big event. Each fire company, as well as the department, proudly celebrates milestones such as this one for the Keyport Engine Company's 50th anniversary in 1939. It was held at Blodgett's restaurant, which was part of the Rollo bus terminal on Highway (Route) 35 at Six Corners in Keyport.

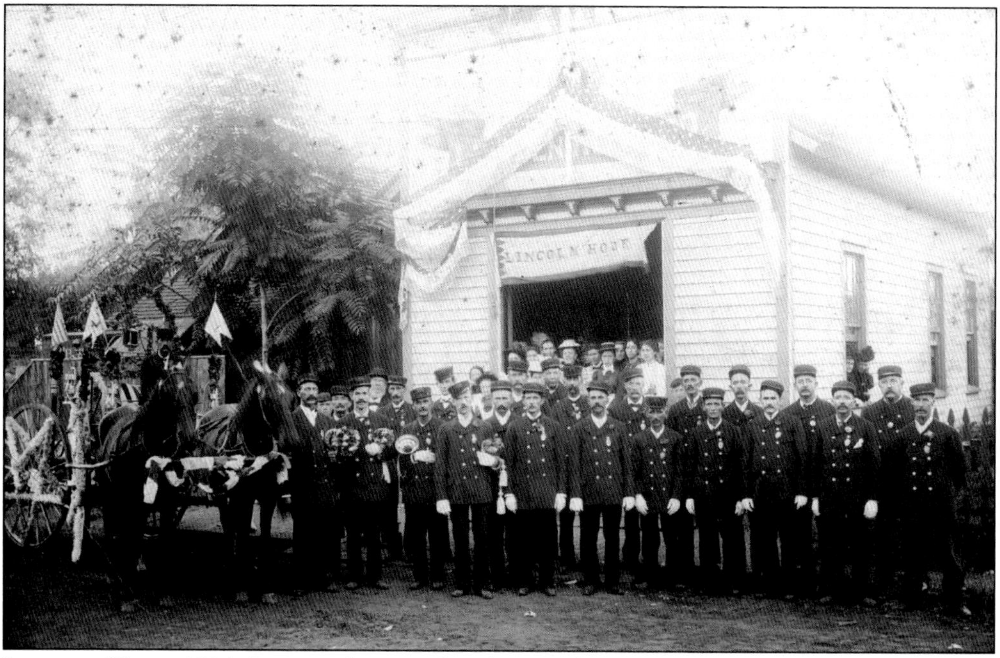

LINCOLN HOSE, 1894. In just a few months, the proud members of this east end fire company had accomplished much. They purchased land, erected this neat one-story firehouse, obtained a fine hose carriage, and outfitted 30 members with uniforms and badges. Many of the men in this photograph are veterans of the Civil War who were responsible for naming the company in honor of their slain president, Abraham Lincoln, in 1893.

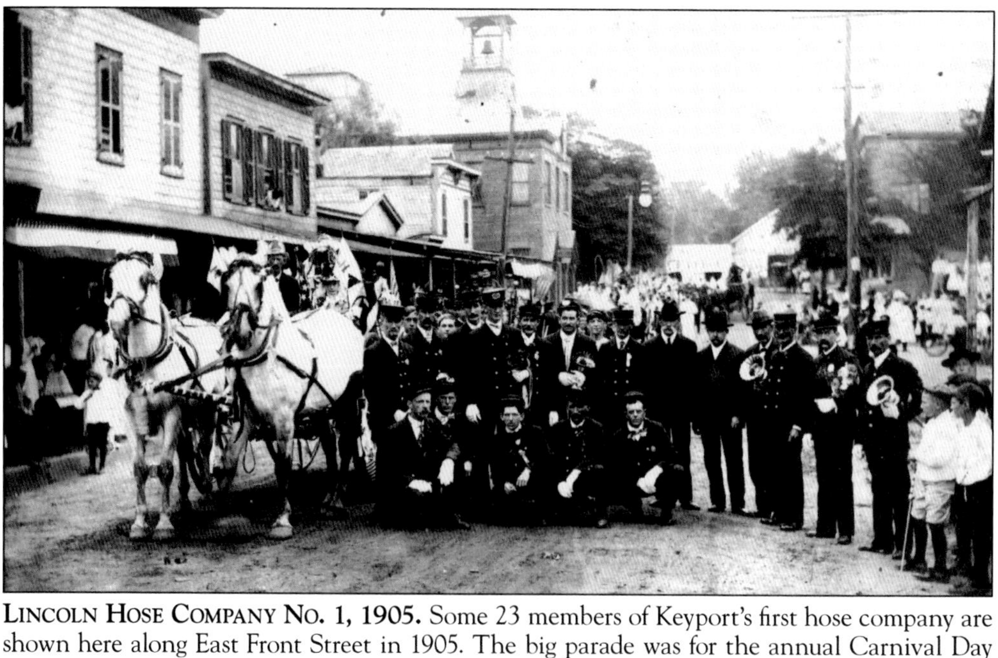

LINCOLN HOSE COMPANY NO. 1, 1905. Some 23 members of Keyport's first hose company are shown here along East Front Street in 1905. The big parade was for the annual Carnival Day or Salt Water Day, which was a tradition that had been passed on since the town's formation. Among the men shown are several Civil War veterans who helped form the Lincoln Hose Company in 1893.

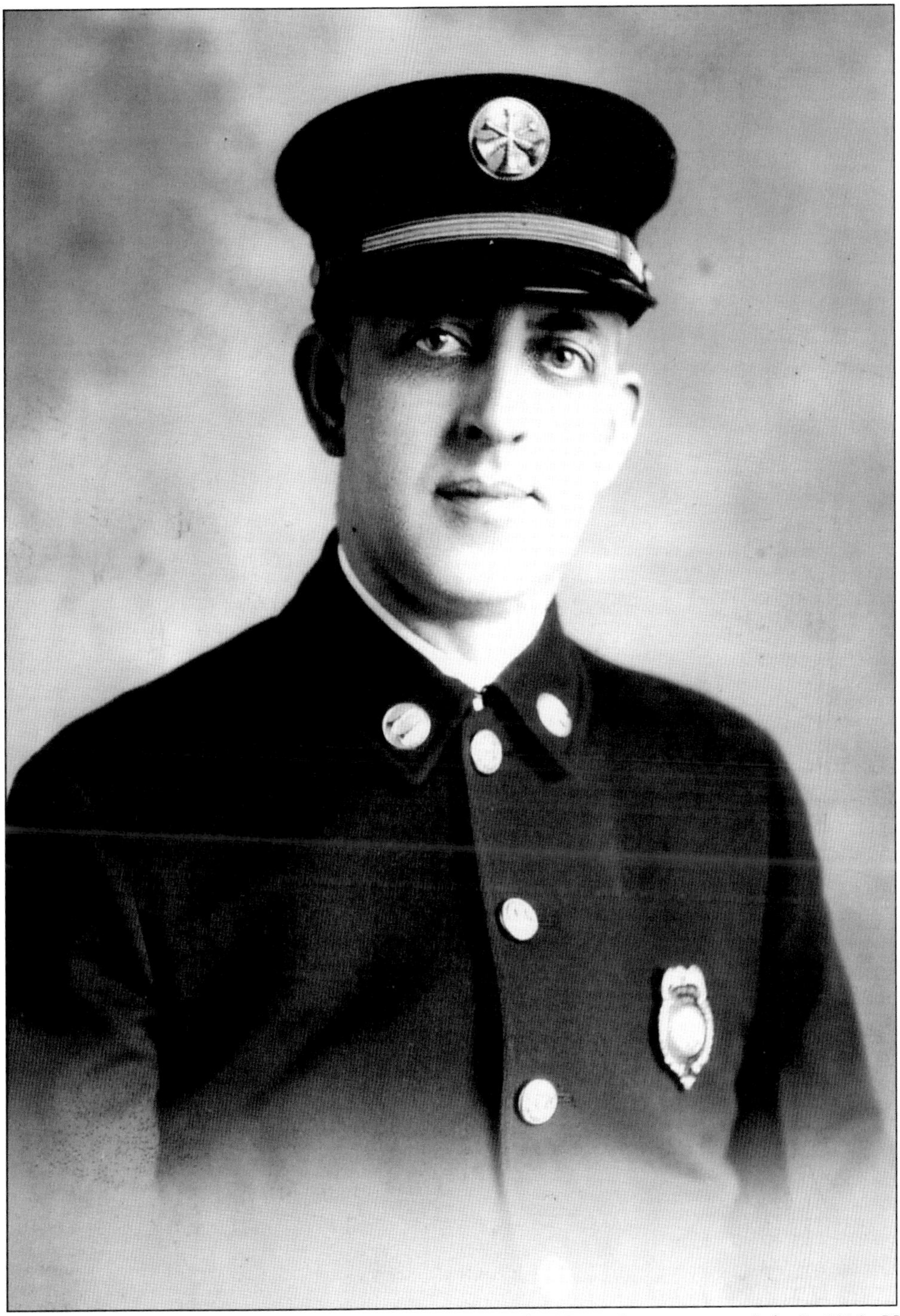

CHIEF BAILEY, 1931. Ernest Bailey joined the Lincoln Hose Company on July 7, 1910. His father, Lafayette Bailey, was a charter member of the company, and several of his brothers also served as firemen. Ernest soon served as foreman of the company and was elected as second assistant chief in 1929. He is best remembered as the father of the Gamewell Fire Alarm system, which he implemented during his tenure as chief. (Courtesy Carolyn Ojamma.)

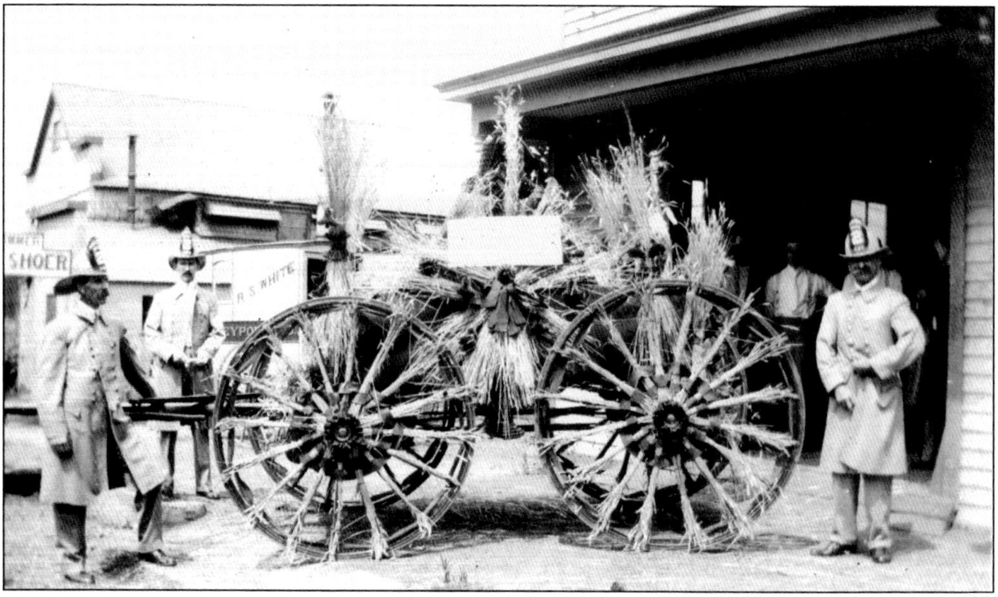

RARITAN HOSE, 1903. Firefighters have taken great pride in decorating their apparatus since the earliest days. Soon the members will hitch up a borrowed horse and parade through town to the delight of all the spectators. In the back, one can see the delivery wagon of R. S. White, a Raritan Hose Company member who often loaned his horse for the purpose of pulling the "old gal" to fires.

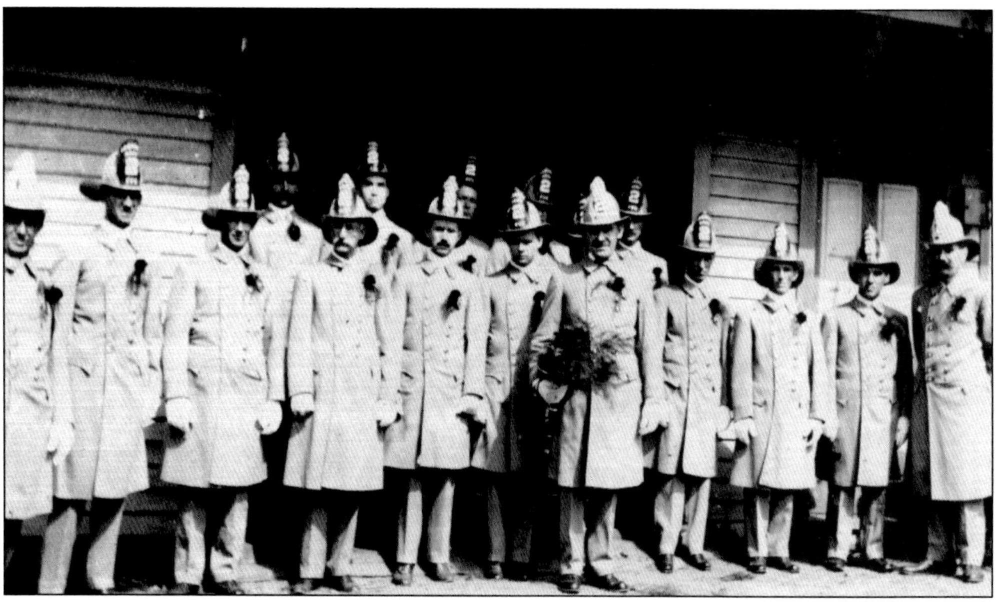

RARITAN HOSE COMPANY NO. 2, C. 1909. Although many of the Raritan Hose Company members were local merchants, all of them shared one common bond—they were all members of the New Jersey National Guard Company G. For many years, all men had to be veterans or former members of the military to become a member of this fine hose company. Included are Elmer Morris, Harry Van Dorn, Frank Stanhope, Richard "Dick" White, George Smith, Ed Cottrell, Tom Burrows, Sy Cramman, George Birch, and Frank Van Brackle.

CHIEF AND MRS. BIRCH, 1937. This unique photograph found in the Raritan Hose Company archives shows Chief George Birch and his wife in formal attire. The Birches resided in the grand Second Empire home built in 1875 by sea captain Samuel Fairchild on First Street. For over six decades, the Birches entertained and welcomed friends and family into their home on First Street. (Courtesy Raritan Hose Company.)

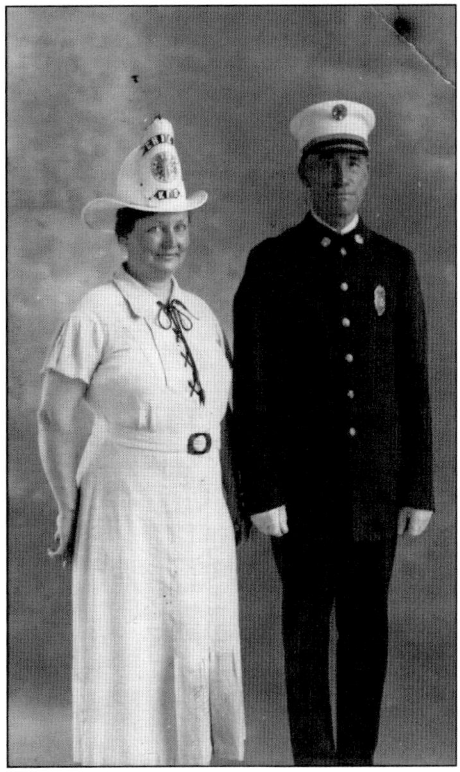

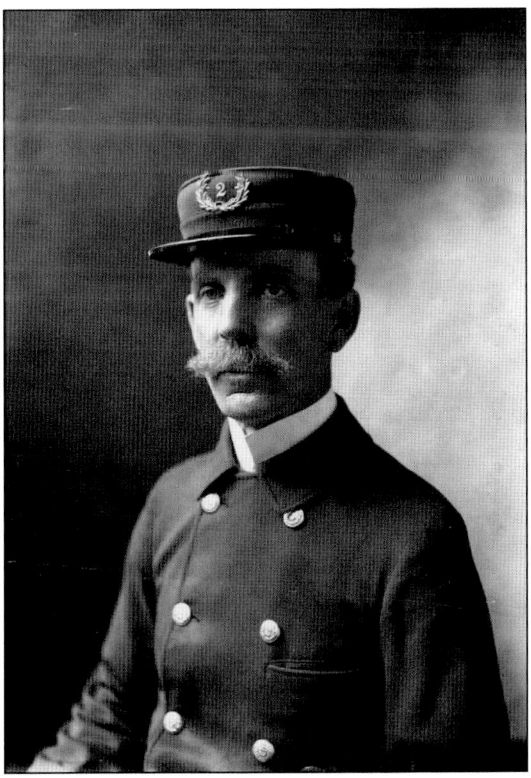

T. WALTER SNYDER, 1899. T. Walter Snyder joined the Raritan Hose Company on September 5, 1898. He was an important Keyport businessman and owned and operated Snyder's Ice Cream parlor at 74 First Street. About 1914, he was elected assistant chief and served as chief of the department from 1918 to 1920. He became inactive in 1925 after nearly 30 years of service. The breast pocket on his uniform is a very unique feature.

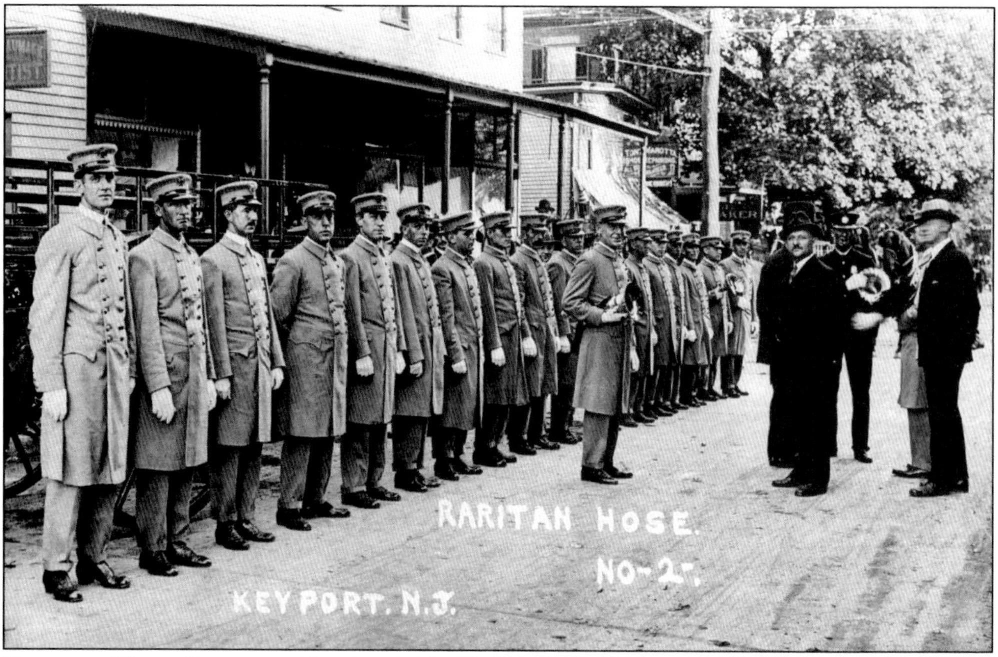

Raritan Hose Company No. 2, 1915. Clad in their dress coats of grey wool, the men of Raritan Hose Company No. 2 are shown along the east side of Broad Street. In this image, the members are being inspected by the mayor, council, and fire chief for company order, proper uniforms, and apparatus cleanliness, as well as overall efficiency. (Courtesy Tom Gallo.)

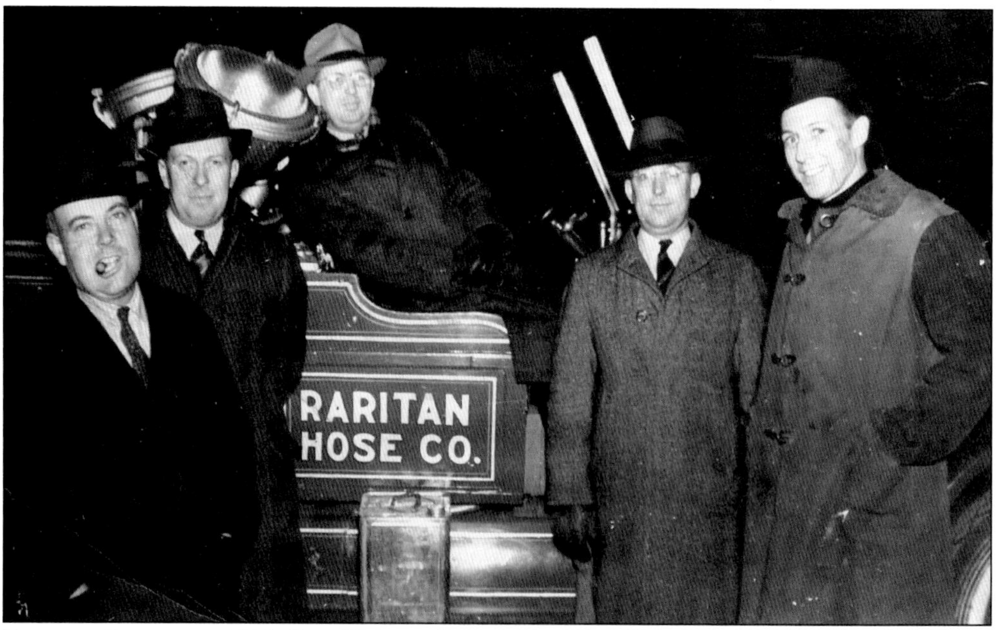

Future Chiefs, c. 1939. Five Keyport firemen pose for the camera alongside Raritan Hose Company's 1926 Graham hose wagon. From left to right are Liberty Hose Company men Ernest Walling, who served as chief in 1943–1944, and Arthur Maitland Walling, who was in chief 1955; Fred Fowler of Raritan Hose Company, chief in 1939–1940; Fred Bronkhurst of Eagle Hose Company, chief in 1951; and Art Collins of Raritan Hose Company, chief in 1960.

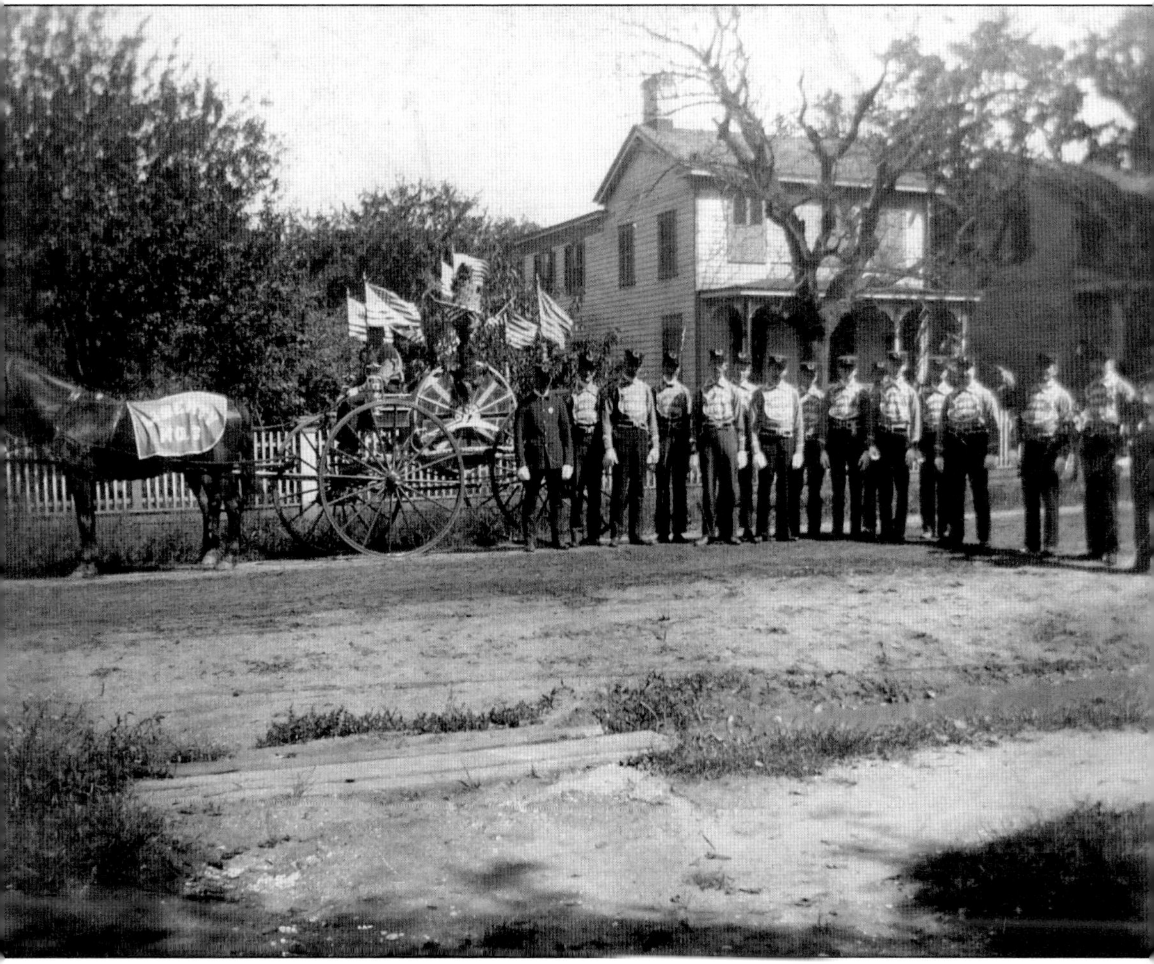

LIBERTY HOSE COMPANY NO. 3, 1900. The entire membership proudly poses with their new uniforms in this 1900 portrait. The company had recently added a second floor to the original Atlantic Street firehouse, as a proper place to entertain the ladies. The company's 1894 Tilton and Cherry hose carriage is in the center.

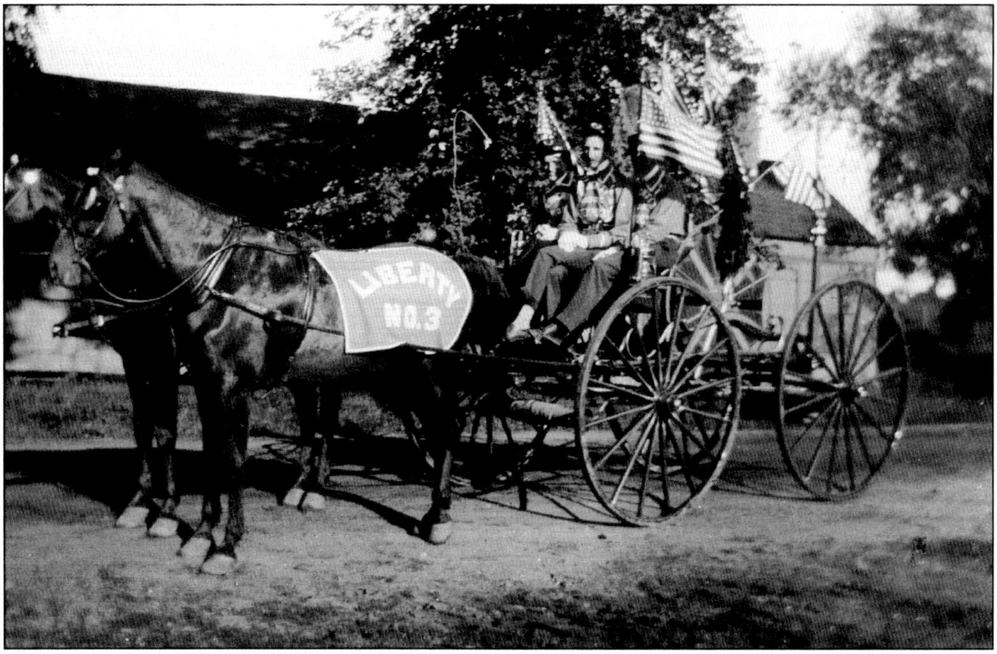

LIBERTY READY FOR A PARADE. Taken from the front of the old Atlantic Street firehouse, this photograph captures Liberty Hose Company around 1903. This is the only photograph that shows a horse blanket with a company name embroidered on it.

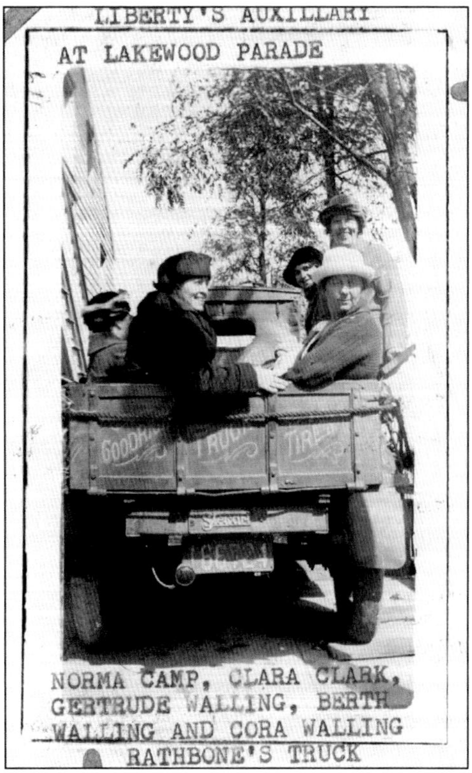

LIBERTY AUXILIARY, C. 1925. Both the Keyport Hook and Ladder Company and Liberty Hose Company had organized ladies auxiliaries by 1930. Their main focus would be to support the company by providing food, drink, and relief at major fires. Here Norma Camp, Clara Clark, Gertrude Walling, Berth Walling, and Cora Walling of Liberty Hose Company parade through Lakewood in the mid-1920s.

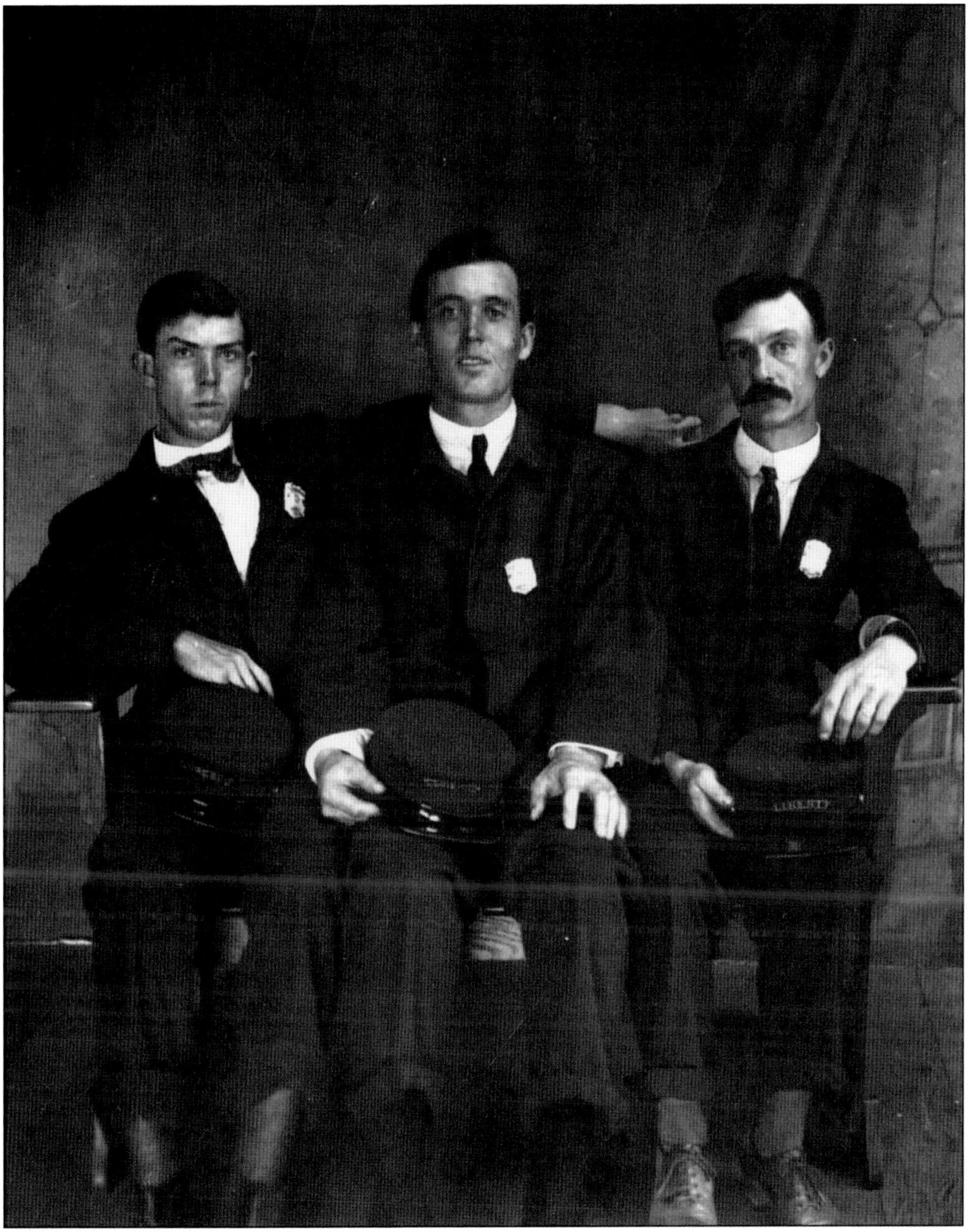

LIBERTY MEN, 1911. These three men gathered for this portrait in the studios of Ferdinand Maag sometime around 1911. Maag Studios was responsible for thousands of portraits of Keyport citizens from about 1886 until the early 20th century. Maag himself was a member of Liberty Hose Company from 1897 until 1905.

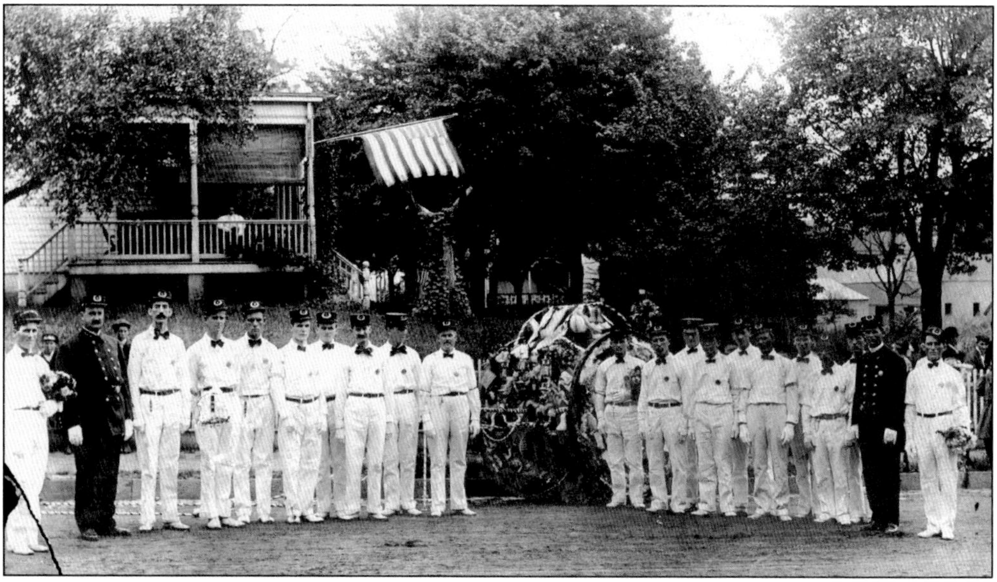

EAGLE HOSE, C. 1908. From 1908 until 1914, the Eagle Hose Company sported these dashing white uniforms. At this time, the company was quartered in a small barn at the rear of the Hopla property on Broadway. In just a few years, Eagle Hose Company would purchase regulation blue firemen's uniforms, which are still worn today.

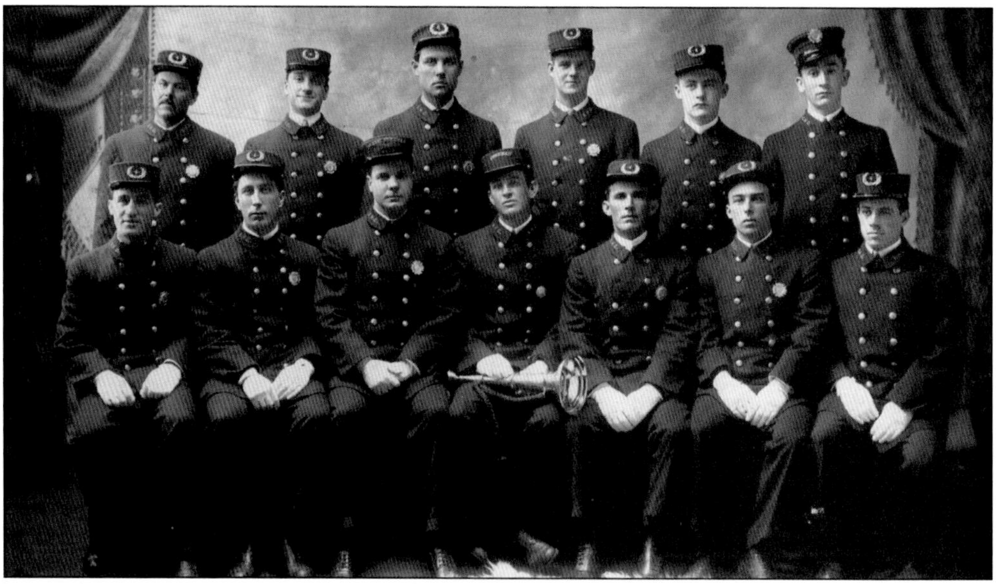

LUKE SLOVER, EAGLE HOSE. Since the department's inception, only two firefighters have given their lives in defense of their country. Luke Slover (first row, far right) joined Eagle Hose Company No. 4 in 1914 and was soon followed by his brother Art the following year. Luke was drafted into the U.S. Army and soon became attached to Company B, 311 Infantry, 78th Division. He took to the trenches in France and was killed in action in 1918. He was awarded the Distinguished Service Cross, which is the second-highest military decoration that can be awarded to a member of the U.S. Army. It is awarded for extreme gallantry and risk of life in actual combat with an armed enemy force. His remains were later returned to Keyport and are interred in the family plot at Green Grove Cemetery.

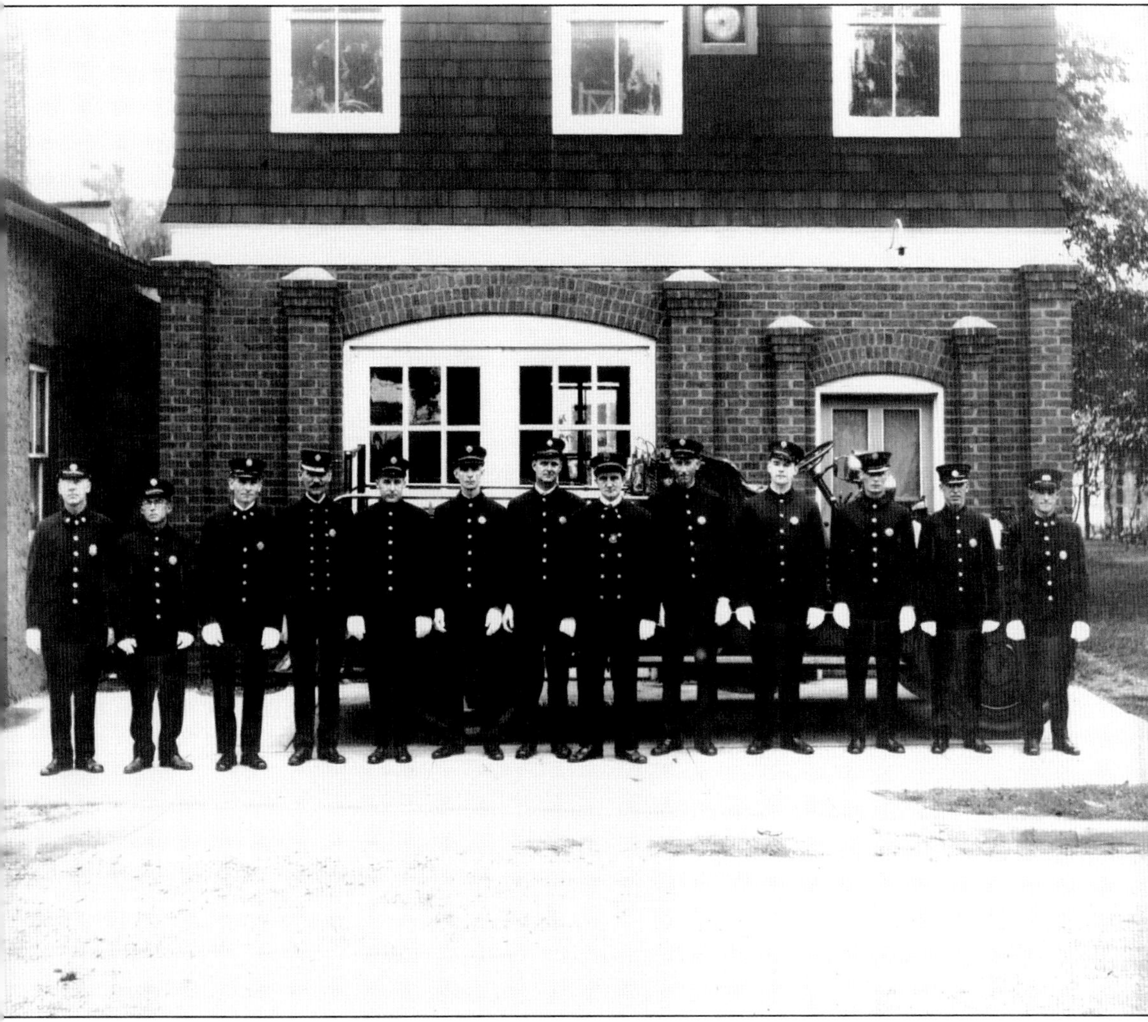

EAGLE HOSE COMPANY, C. 1925. Eagle Hose Company completed its new firehouse in 1912 at 95 Broadway. Company members are standing against the 1922 Federal hose truck, which was the company's third apparatus. Included are James H. Cadoo, Michael Jones, Harry Poling, Pete Morris, Chief William Morris, Fred Bronkhurst, Ed Walters, Jack Haley, and William Waitt Sr. Interestingly, the Cadoo family has had an active member in this company since 1921.

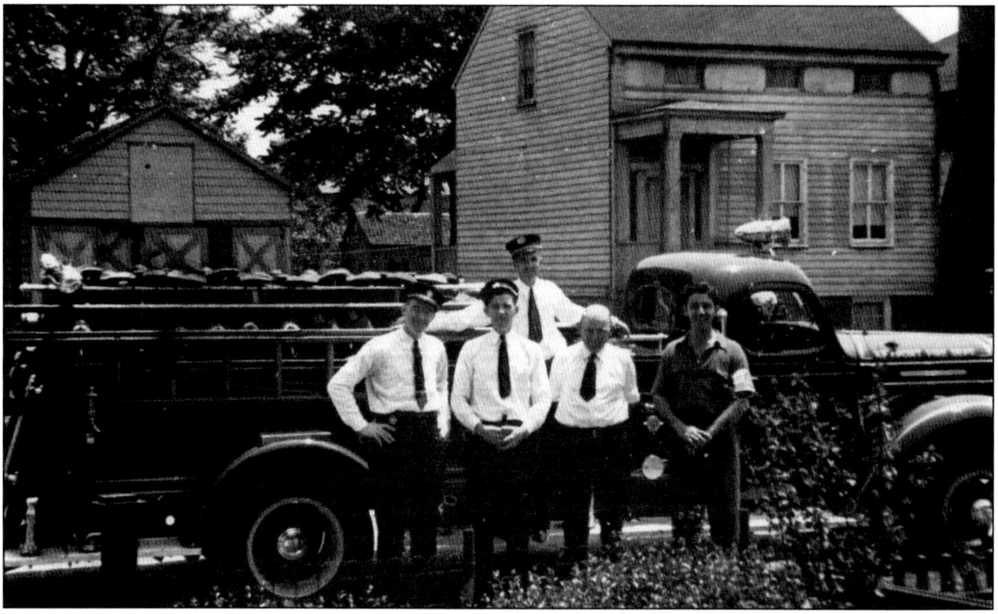

EAGLE HOSE, 1942. This interesting photograph captures a group of Eagle Hose Company members waiting for a parade on Broadway during World War II. However, in the background what appears to be an old garage is actually the first firehouse of the company, which was rebuilt in 1907 and served the company for six years until the new firehouse was built down the street.

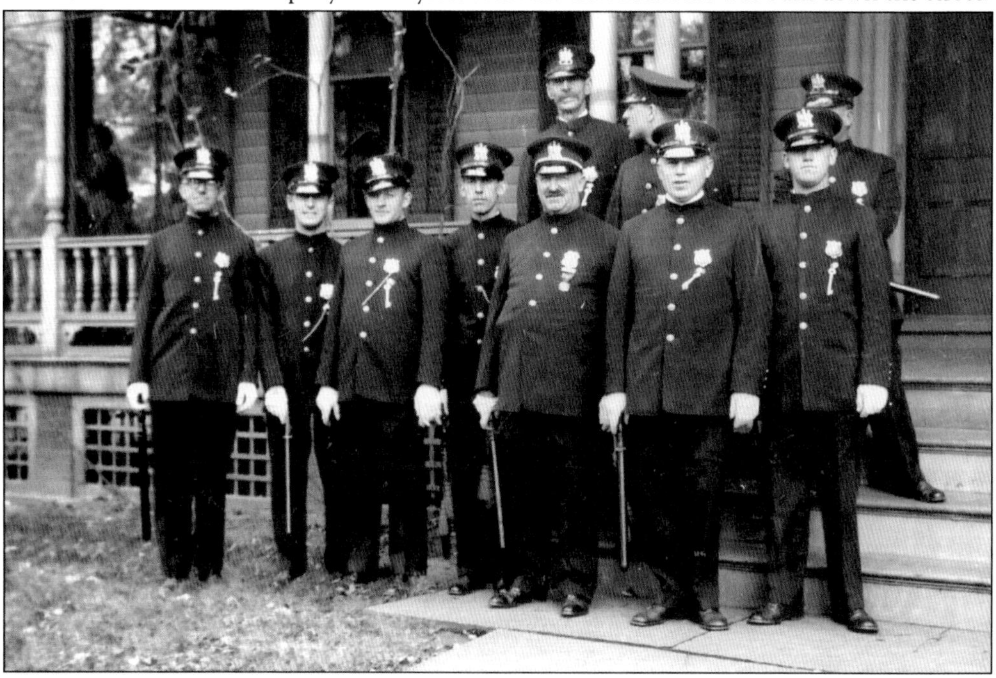

FIRE PATROL, 1927. By 1912, the old fire patrol had been replaced by the Board of Fire Wardens. Their job was threefold: keep order at fire scenes, protect private property, and arrest violators. By the time of the 50th anniversary of the KFD, each fire warden had been issued a wooden nightstick and a key to the town jail, just in case young Johnny crossed the fire line one too many times. (Courtesy Keyport Fire Patrol.)

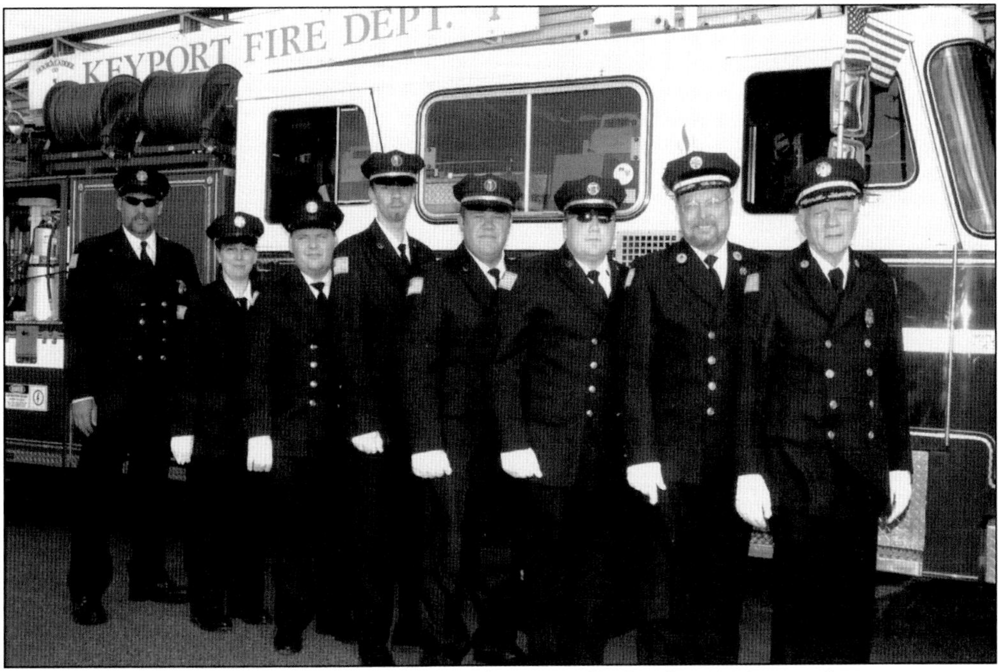

Hook and Ladder Company No. 1. Standing proudly for the 2006 annual inspection are, from left to right, Jim McTernan (coauthor), Jacqueline Kovacs-Olsen, Daniel Fox, Richard Leswing VII, David Quinn, Joseph Walling, and ex-chiefs Roger Benedict and Harry M. Aumack II.

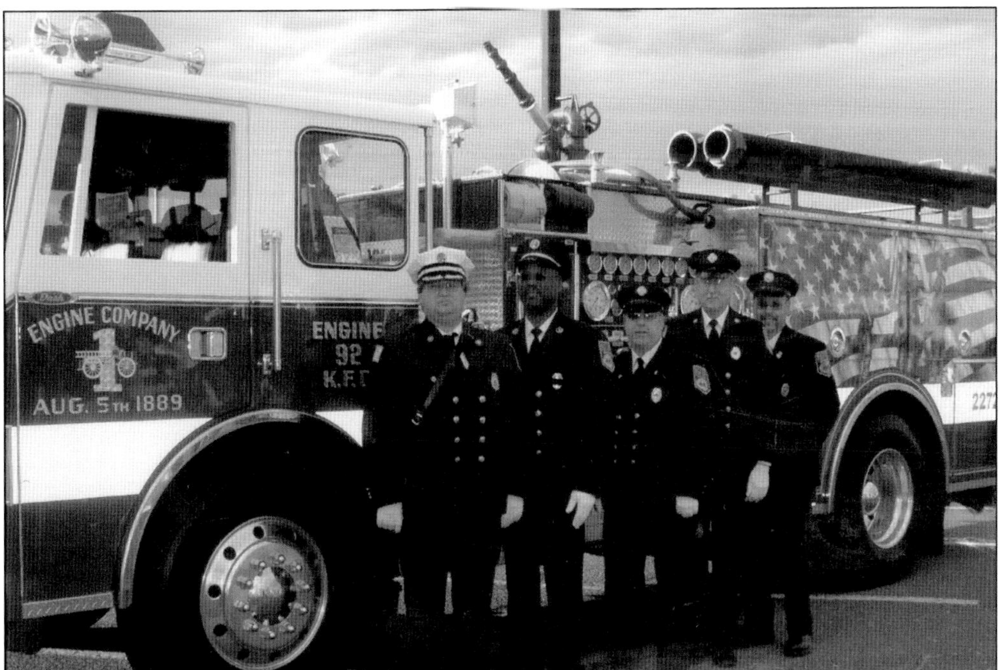

Keyport Engine Company No. 1. Pictured here are, from left to right, Chief David Olsen, Capt. James Lawson, George Strang, Robert McCleod, and Jose Mateo.

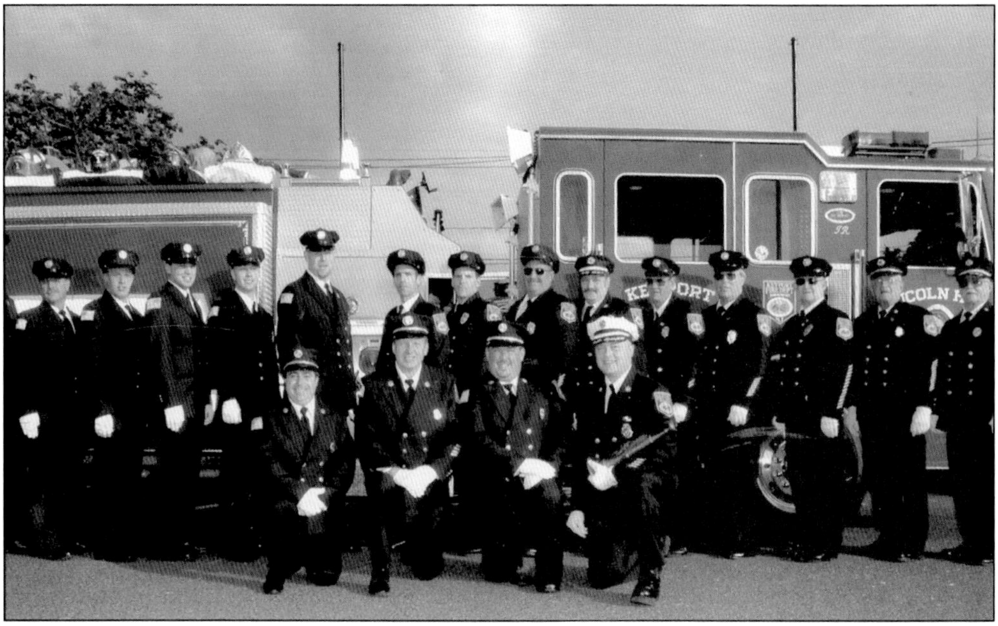

LINCOLN HOSE COMPANY NO. 1. Seen here are, from left to right, (kneeling) Lt. Lou Troisi, Lt. Robert Aumack, Capt. Thomas Grabowski, and Chief Tim Regan (coauthor); (standing) Kieran Gormley (out of view), Charles McCue, Phil Santiago, Joseph DeSandolo, Daniel Regan, John Condon, Christopher Poling, Robert Poling, Frank Grabowski III, ex-chief Pete Rumsey, John English, John Pangborn, James Poling, and ex-chiefs Gene Regan and Robert Poling.

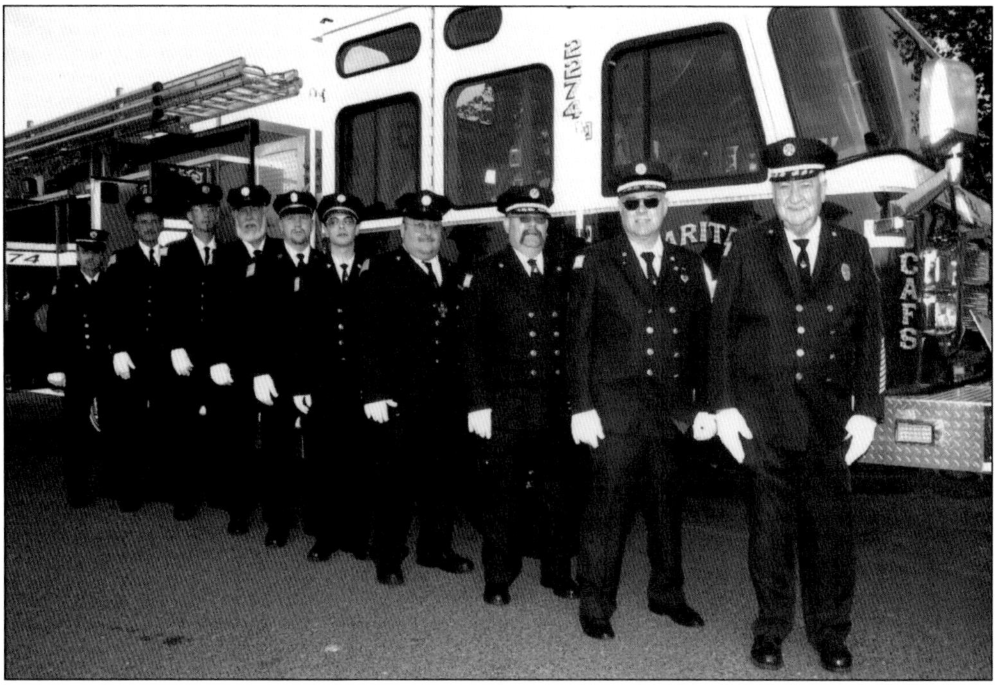

RARITAN HOSE COMPANY NO. 2. From left to right are Robert Baird, Robert Ciecierski, Michael Ciercierski, Robert Ciecierski Sr., Matthew Gajewski, William Jeandron, department chaplain Raymond Kilroy, and ex-chiefs Roy Jimenez, William Jeandron, and John Jones.

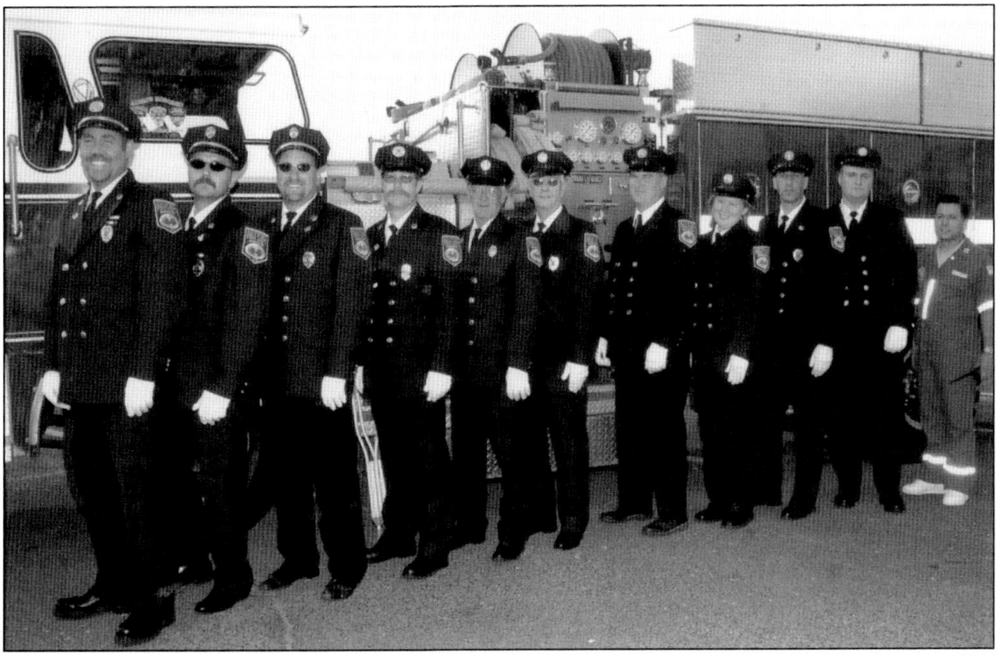

Liberty Hose Company No. 3. Seen here are, from left to right, Capt. Chris Alcott, Lt. Dallas Smith, Lt. Joseph Pinto, John McCleaster, George Sprague, Jack Alcott, John Meyer, Kaitlin McCann, Scott K. Eack, Roman Doegykh, and ex-chief Ken Krohe.

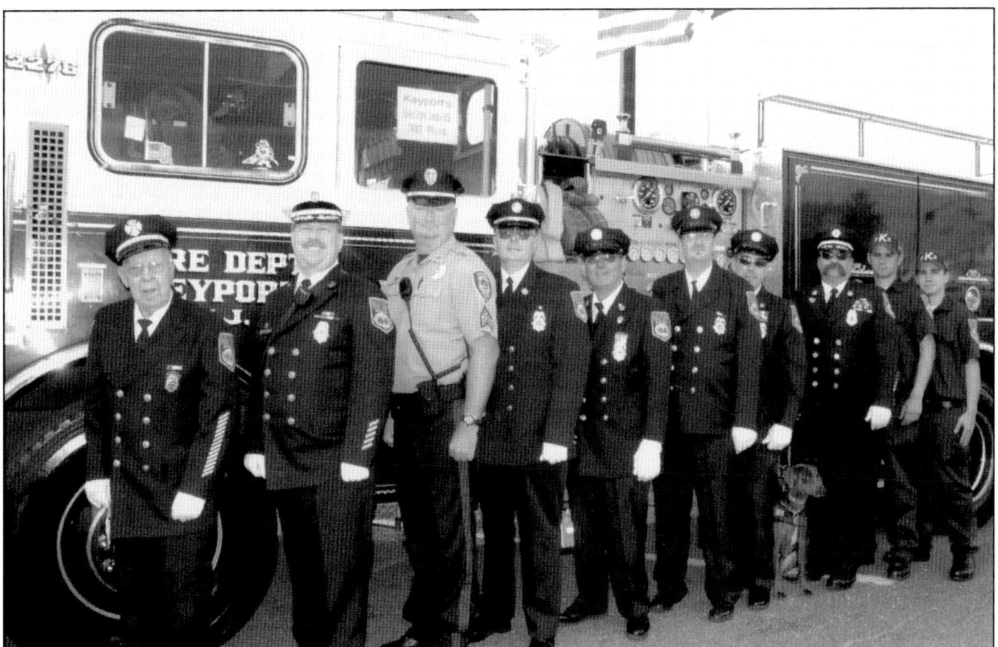

Eagle Hose Company No. 4. Seen here are honorary chief Meldon "Babe" Stultz, Chief William Larkin, Capt. Rick Ely, Lt. Gerald Greene, Terry Groffie, Joseph Corson, ex-chief Anthony Vecchio, Nicholas Vecchio, and Jacob Vecchio.

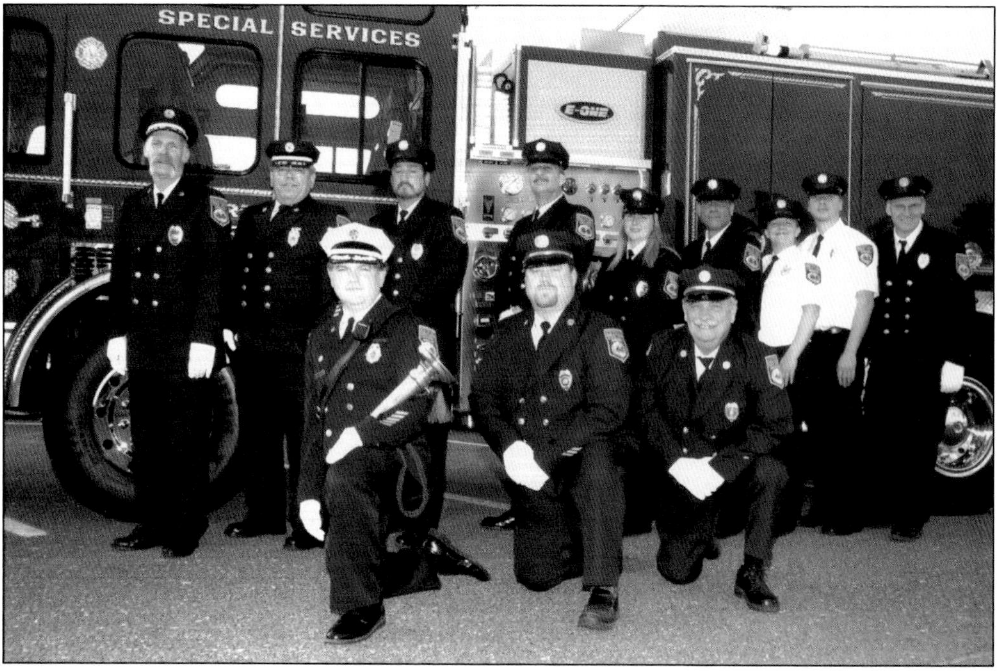

FIRE PATROL. Seen here are, from left to right, (first row, kneeling) Chief William Vaughn, Capt. Charles Pfleger, and Lt. Nick Devincenzo; (second row) ex-chief Ken Goble, ex-chief Lawrence Stonerock, Joseph Konish, Tom Gallo (coauthor), Elizabeth Trahim, Charles Raftery, Julie Ann Csik, Gary Wolchesky, and Jay Gordon.

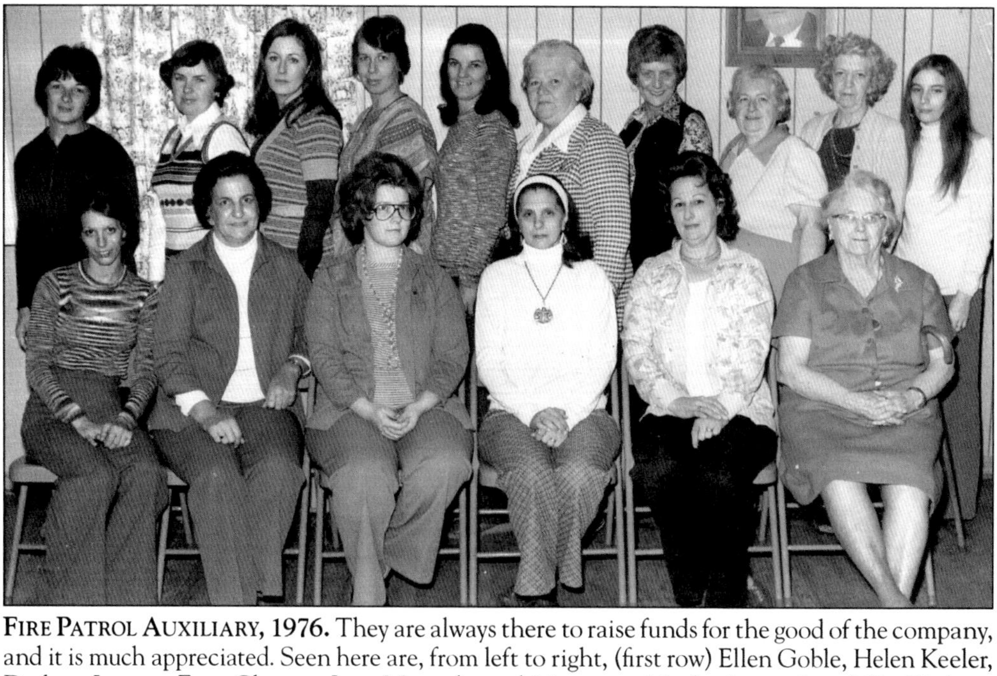

FIRE PATROL AUXILIARY, 1976. They are always there to raise funds for the good of the company, and it is much appreciated. Seen here are, from left to right, (first row) Ellen Goble, Helen Keeler, Darlene Larson, Fran Clayton, Jane Muench, and Margarate Nagle; (second row) Pat Trolian, Linda Rudolph, Peggy Jutie, Janet Richards, Sandy Siegle, Lee Trolian, Ronnie Shaw, Violet Tice, Valerie Ferry, and Louise Conrad.

Three

Apparatus and Firehouses

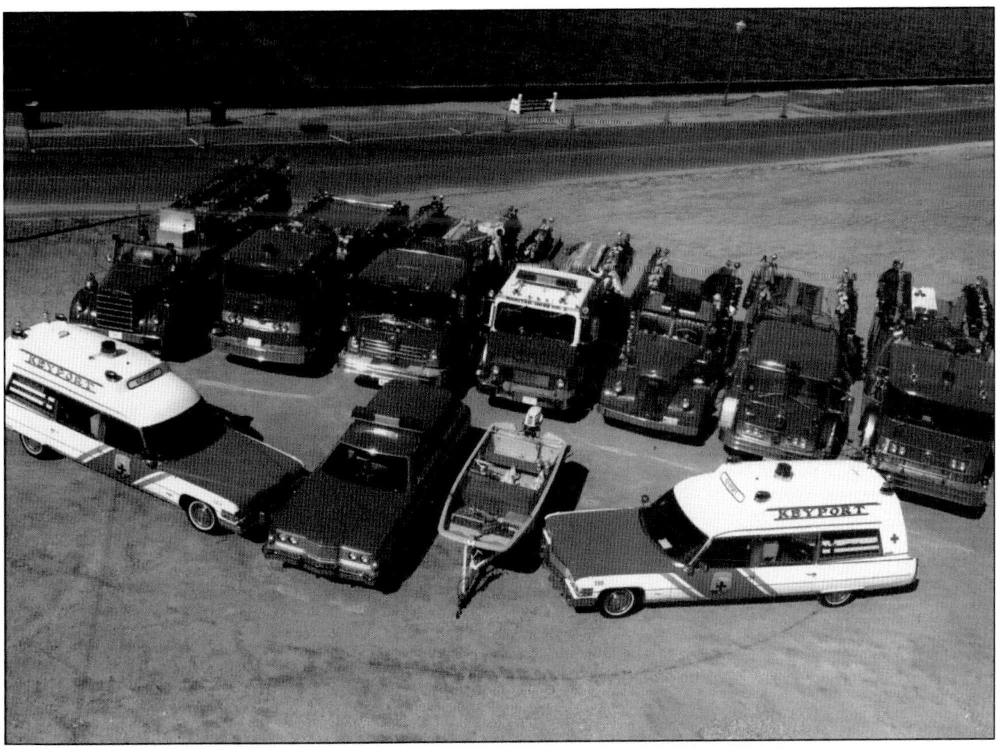

Fire Department, 1977. Every time an apparatus change is made, a new department portrait is taken for posterity. Here the fleet sits along American Legion Drive during the 100th anniversary year of the department. Today only the open-cab Ward LaFrance pumper remains and serves as a parade piece for the Liberty Hose Company.

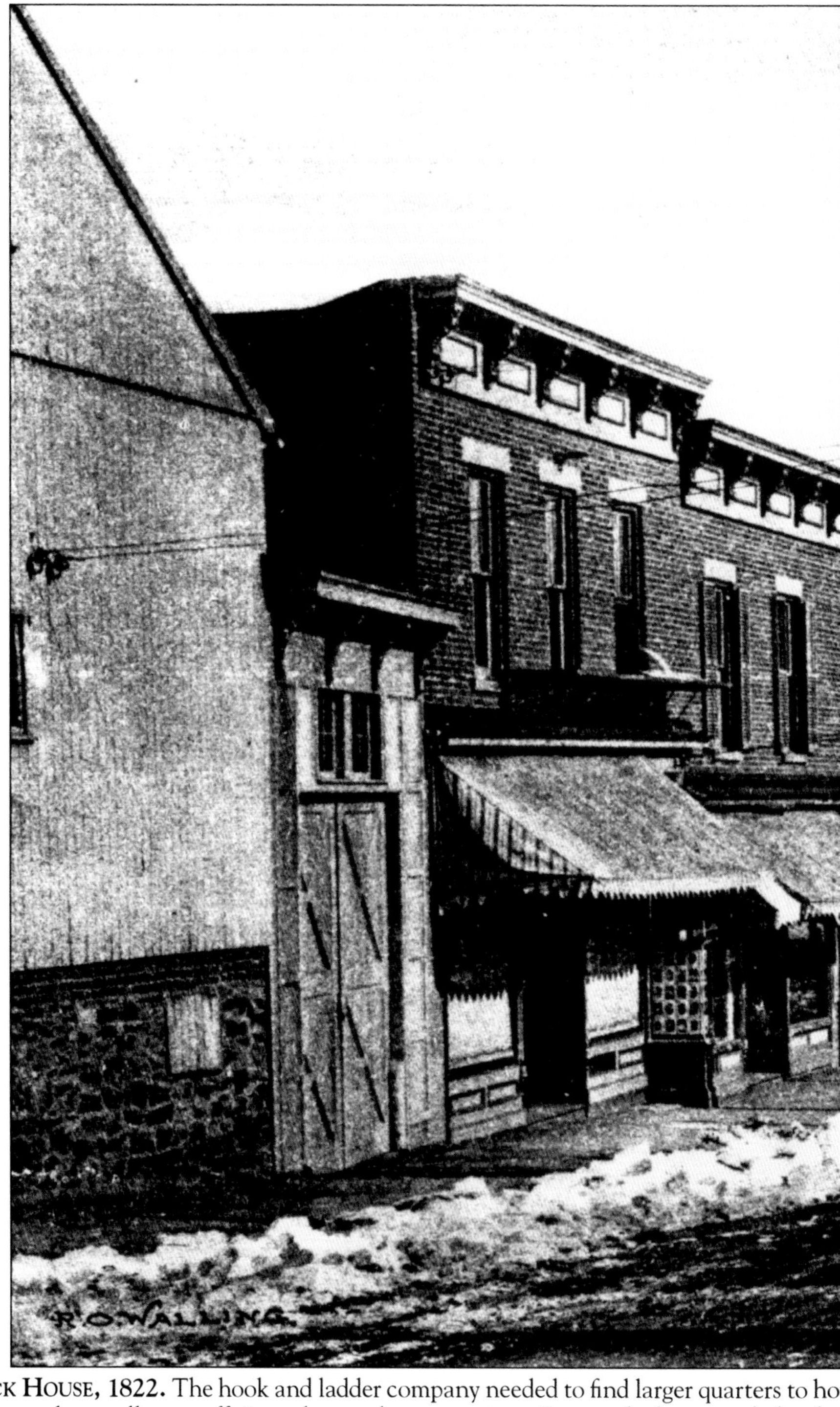

TRUCK HOUSE, 1822. The hook and ladder company needed to find larger quarters to house the ladder truck as well as a sufficient place to have meetings. Soon a deal was made by the owner of the old bowling alley on East Front Street near the Mansion House stables, and the company moved in by April 1882. It served for just a few short years, when the company moved a half block away to the new town hall. (Courtesy Jim McTernan.)

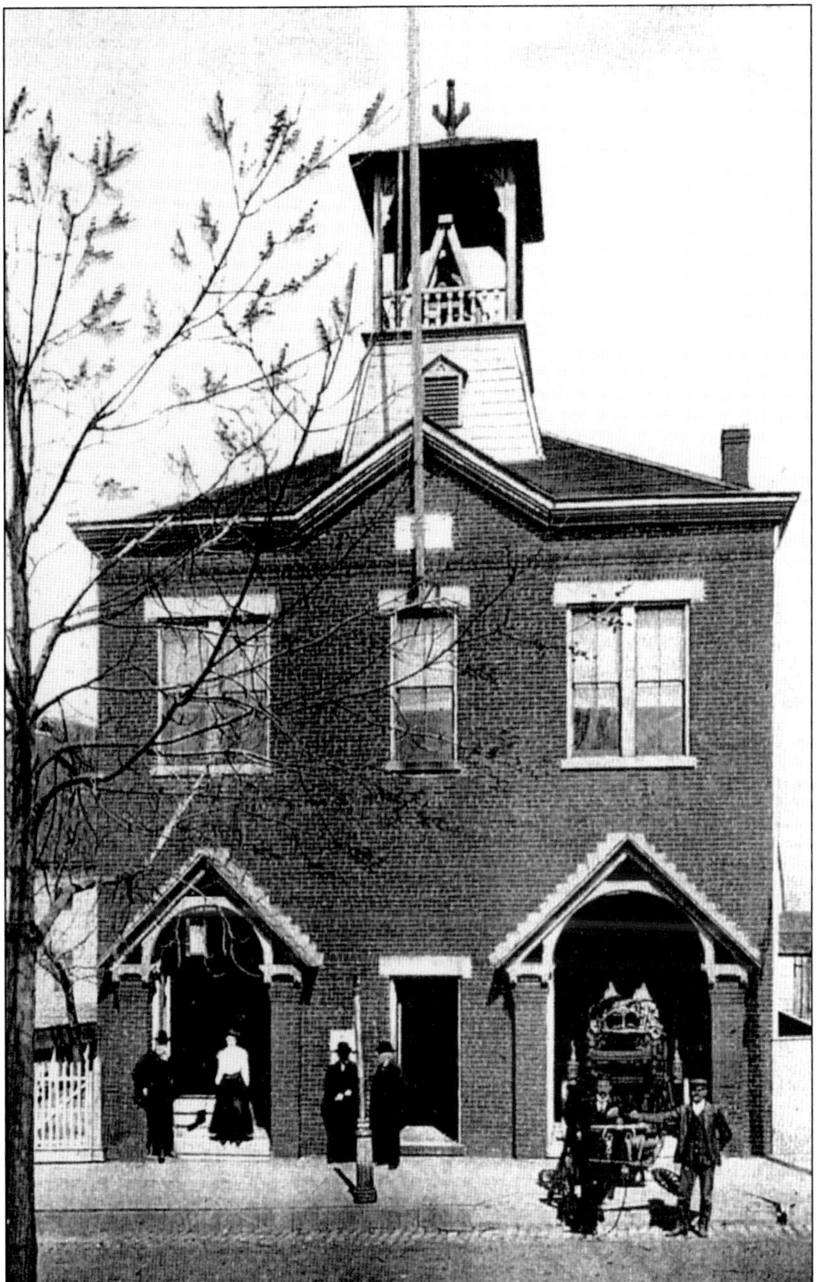

TOWN HALL AND FIREHOUSE. Keyport's third location for a firehouse came with the completion of this building in 1885. Located on East Front Street, the imposing brick building had room for all town offices, a jail, council chambers, and the hook and ladder wagon. The building was first occupied in 1886, and in 1887, the fire bell was placed in the tower. The bell was cast at the McShane Bell Foundry in Baltimore and weighs over 1,000 pounds. In 1963, it was removed from the tower and became the centerpiece of the Keyport Fireman's Monument. In 1952, the town fathers relocated, and it then became property of the "Hooks," solely as a firehouse until 1986. (Courtesy Jim McTernan.)

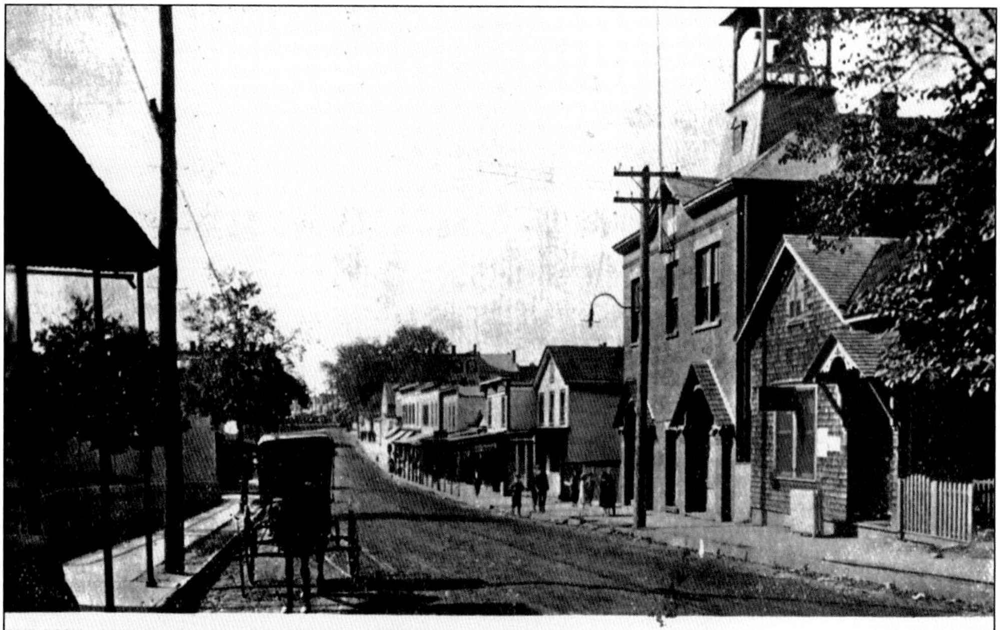

TOWN HALL LOOKING WEST. This postcard view gives a nice indication of the old neighborhood surrounding the 1885 town hall. By 1880, this section of East Front Street had begun to be built up and was located on the site of the Dudley Canning Factory, which was destroyed by fire on February 27, 1877, just a few months before the great fire. (Courtesy Jim McTernan.)

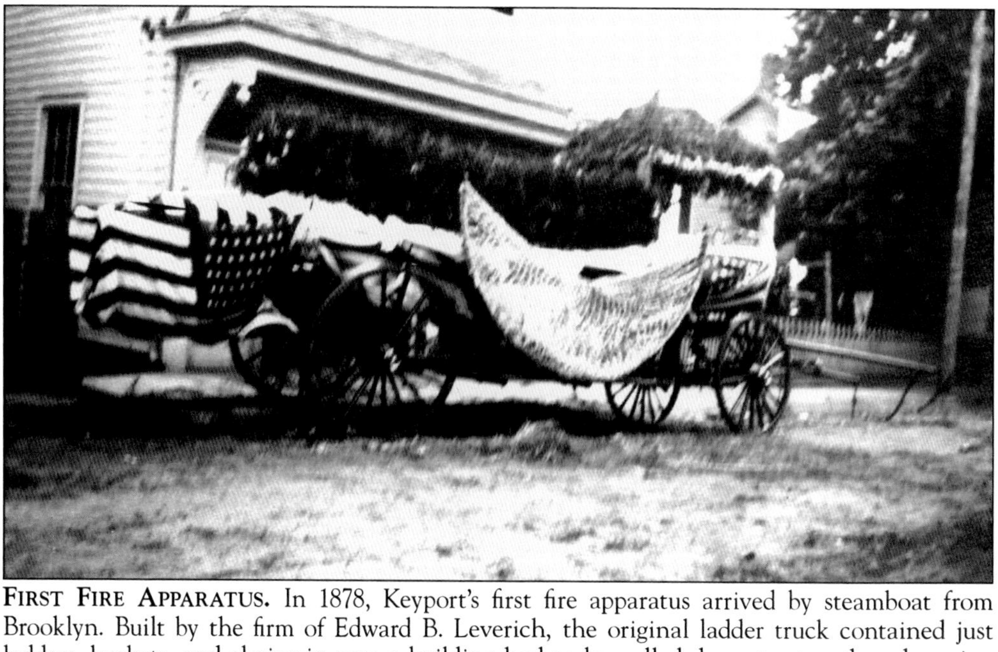

FIRST FIRE APPARATUS. In 1878, Keyport's first fire apparatus arrived by steamboat from Brooklyn. Built by the firm of Edward B. Leverich, the original ladder truck contained just ladders, buckets, and chains in case a building had to be pulled down to stop the advancing flames. In 1909, the truck was replaced and turned over to the Lincoln Hose and Hook and Ladder Company, where it served for an additional six years on the east side of town.

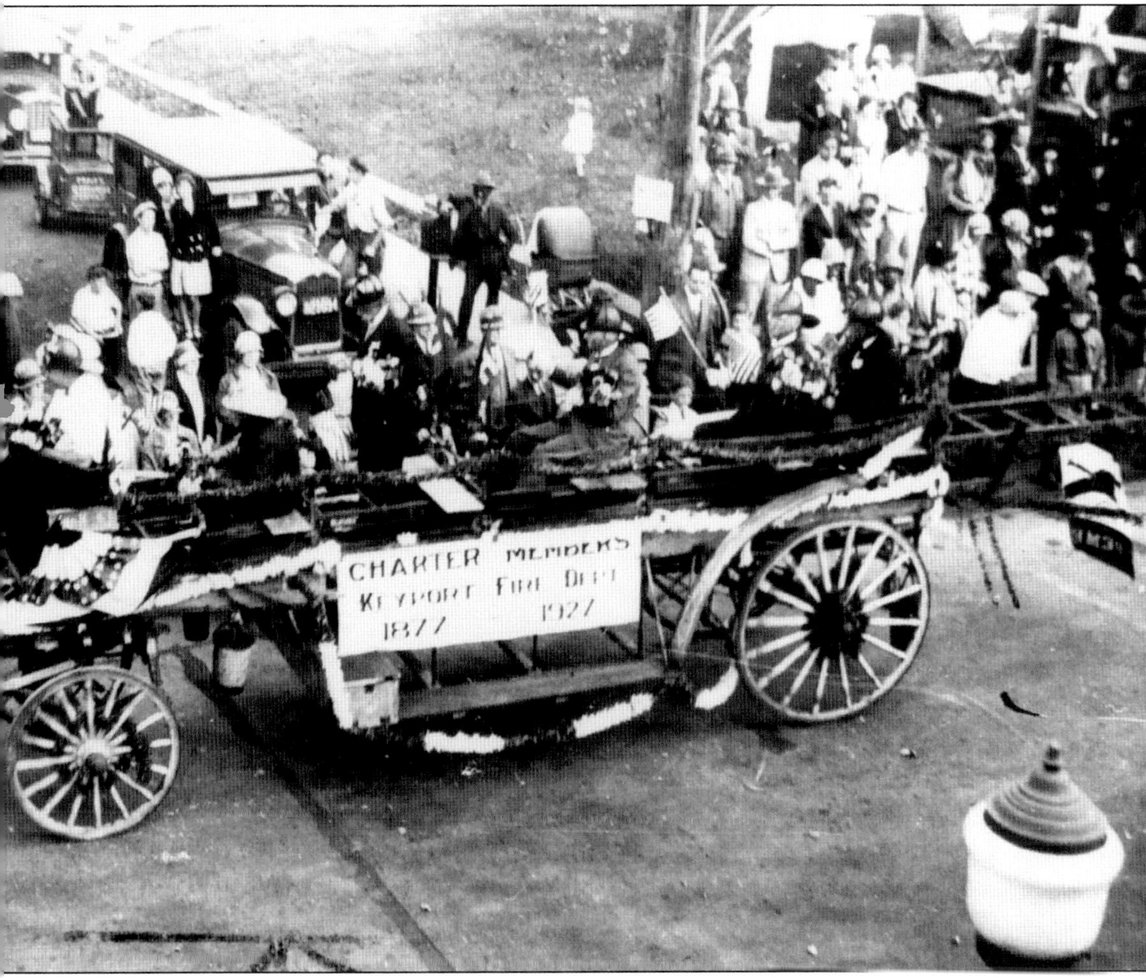

KFD's 50th Anniversary, 1927. The KFD threw a gala event in October 1927 for its half-century birthday. Five of the original 100 charter members gathered to ride on the old ladder truck that day.

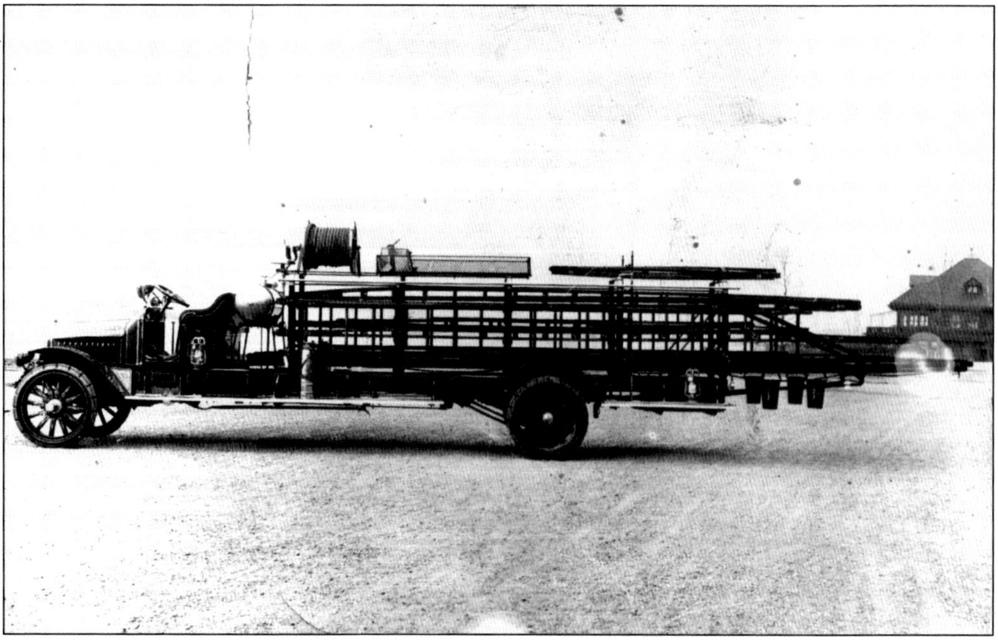

FIRST MOTORIZED LADDER TRUCK, 1920. It was a rainy night when the Broadway Theatre burned, and 10 men pulled "man killer" all the way to Brown's Point to battle a blaze. The ladder was the last to arrive, as every other company except Keyport Fire Patrol was already motorized. On that night, the men decided that manpower was a thing of the past and soon purchased this Mack city service truck, which served until 1934.

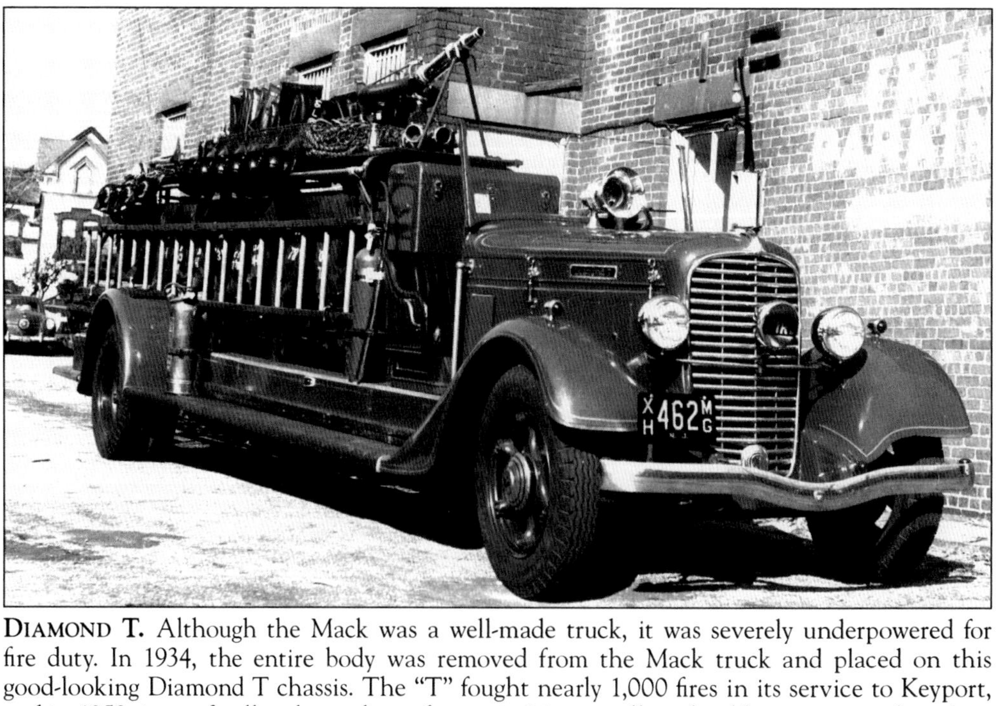

DIAMOND T. Although the Mack was a well-made truck, it was severely underpowered for fire duty. In 1934, the entire body was removed from the Mack truck and placed on this good-looking Diamond T chassis. The "T" fought nearly 1,000 fires in its service to Keyport, and in 1958, it was finally relegated to a farmer in Morganville to haul hay, grain, and produce to area markets—albeit at an acceptably slower pace.

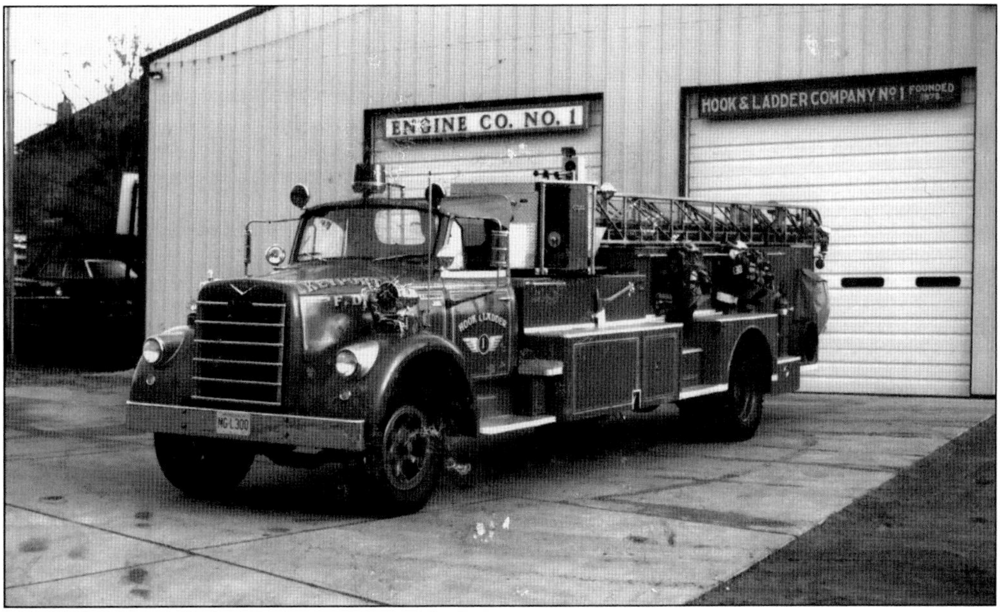

FIRST AERIAL LADDER. In 1958, the Seagrave Corporation of Columbus built this 65-foot aerial on an International 190 chassis for Keyport Hook and Ladder Company. It became the first aerial in this part of Monmouth and Middlesex Counties and fought every major blaze in the area in its 30-year career. It was Keyport's first aerial ladder and the last apparatus to be delivered with a semi-open cab.

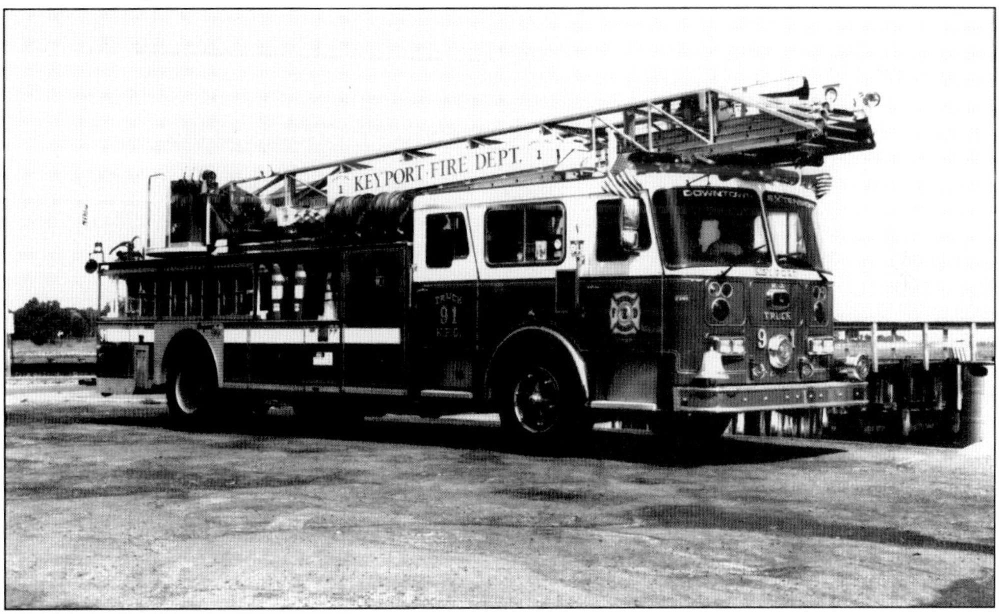

THE 1988 SEAGRAVE. After nearly a decade of planning, the town authorized funding for a new aerial ladder truck. Once again, the Seagrave Company was selected and turned out this fine-looking, 100-foot aerial ladder. It became the first fully enclosed truck in the area and was delivered with a full complement of wooden ground ladders, as well as a Detroit diesel engine. It was the first time since 1921 that a pump and booster line was included on a Keyport ladder truck.

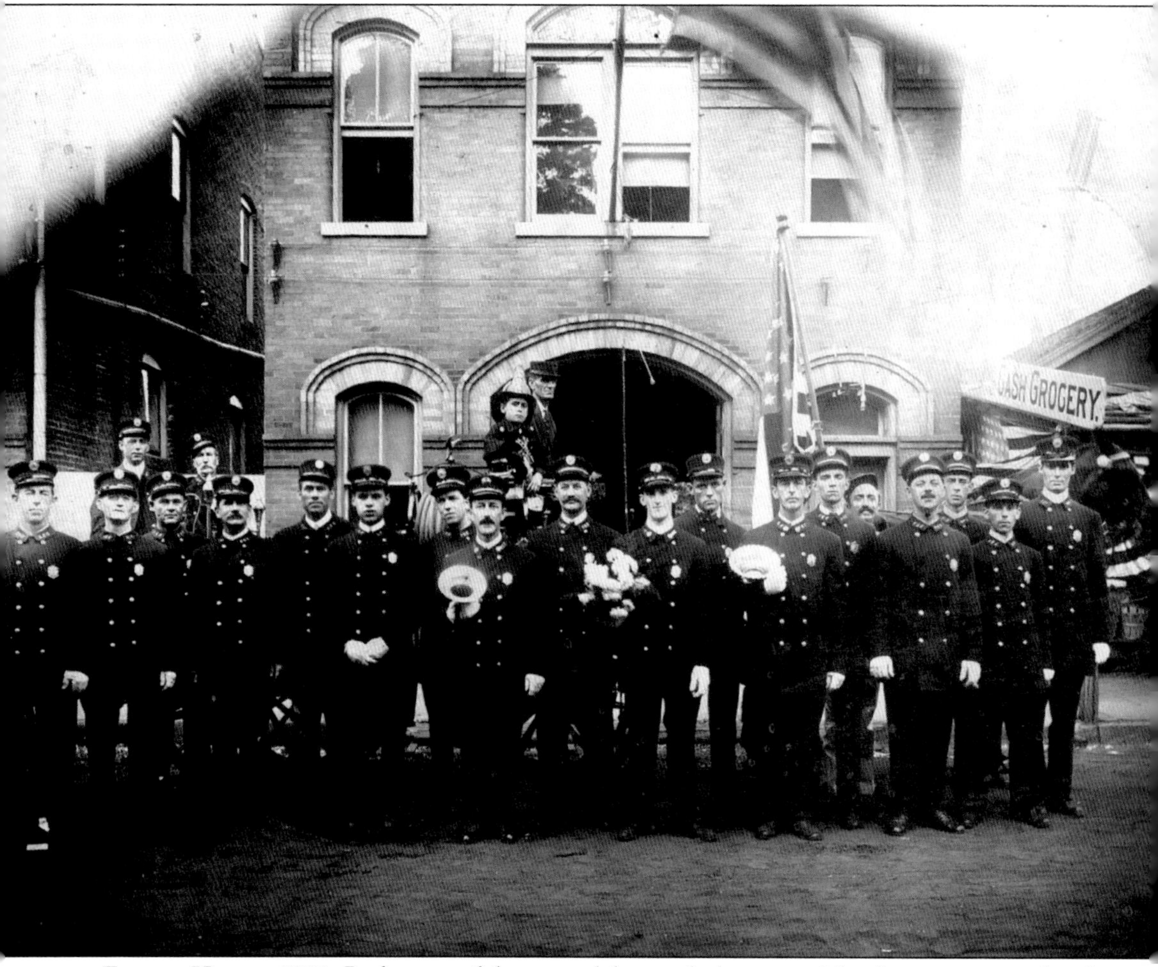

ENGINE HOUSE, 1891. By far, one of the most elaborate firehouses in New Jersey was completed for Engine Company No. 1 on Main Street in 1891. After just about a year of construction, Engine Company No. 1 moved here from First Street and settled into its new home. Here the membership gets ready to step off for an annual inspection parade around 1910.

ENGINE HOUSE. Keyport Engine Company was organized on August 5, 1889, by the membership of the Keyport Athletic Club. When the Holloway chemical wagon arrived in late 1891, it became Chemical Engine Company No. 1. The "chemical" name was used until 1919, when the company received its American LaFrance pumping engine.

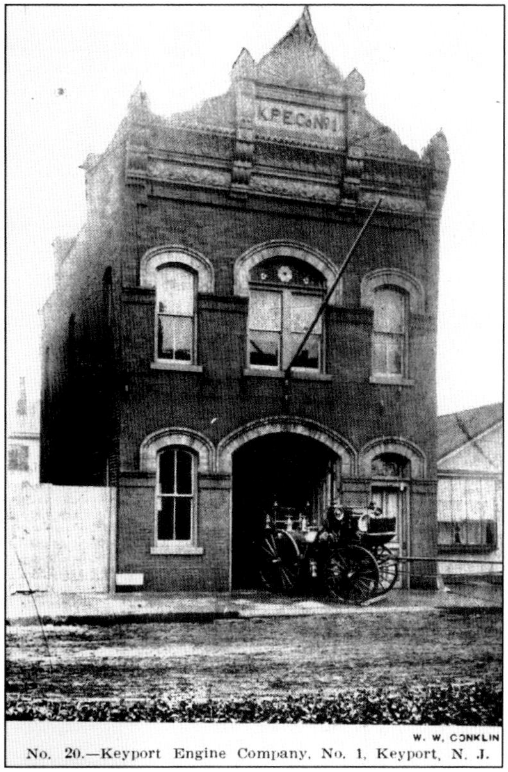

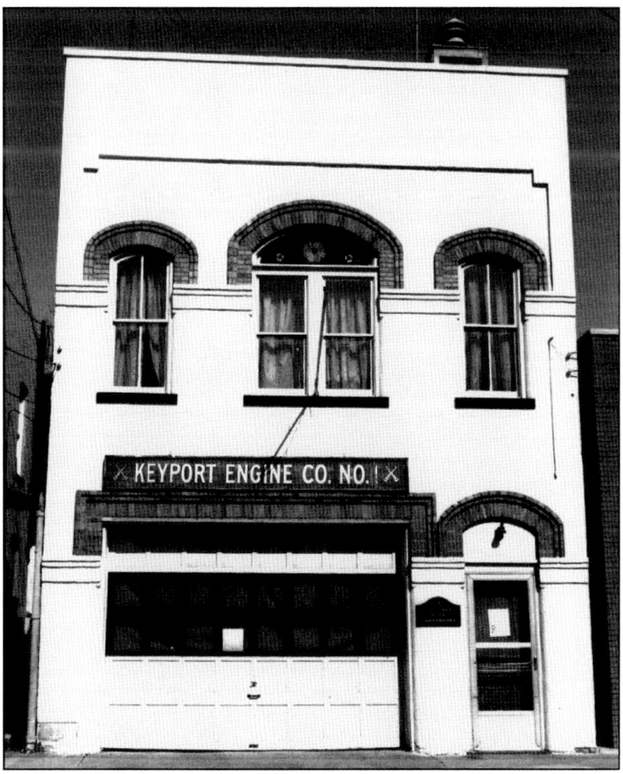

ENGINE HOUSE, SPRING, 1972. During the late 1930s, the firehouse was modernized by removing the ornate tin cornice and widening the door for modern apparatus. By 1928, the building was home to both the American LaFrance and the Keyport First Aid Squad ambulance. The Keyport First Aid Squad remained here for over 20 years, until it moved to quarters on Broad Street. In 1986, Keyport Engine Company sold the building and moved into shared quarters with the Hooks at 34 First Street.

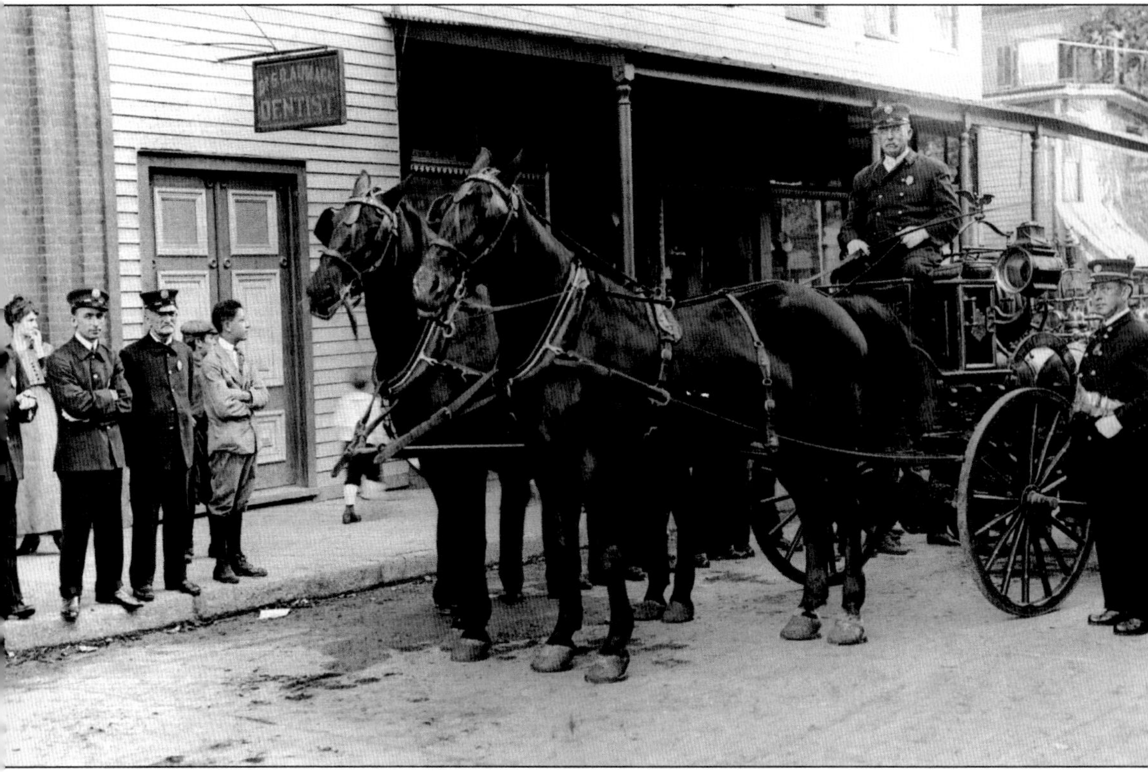

HOLLOWAY CHEMICAL WAGON, 1915. Without a waterworks or hydrants in place, the Keyport Engine Company had no choice but to purchase this chemical truck in 1891. The wagon was equipped with two 35-gallon tanks, one containing water and the other containing a solution of sodium bicarbonate. When a fire stream was needed, a mixing valve would be opened, forcing water through the hose by chemical chain reaction. With just 70 gallons of firefighting water aboard, the little rig saved countless buildings throughout the town.

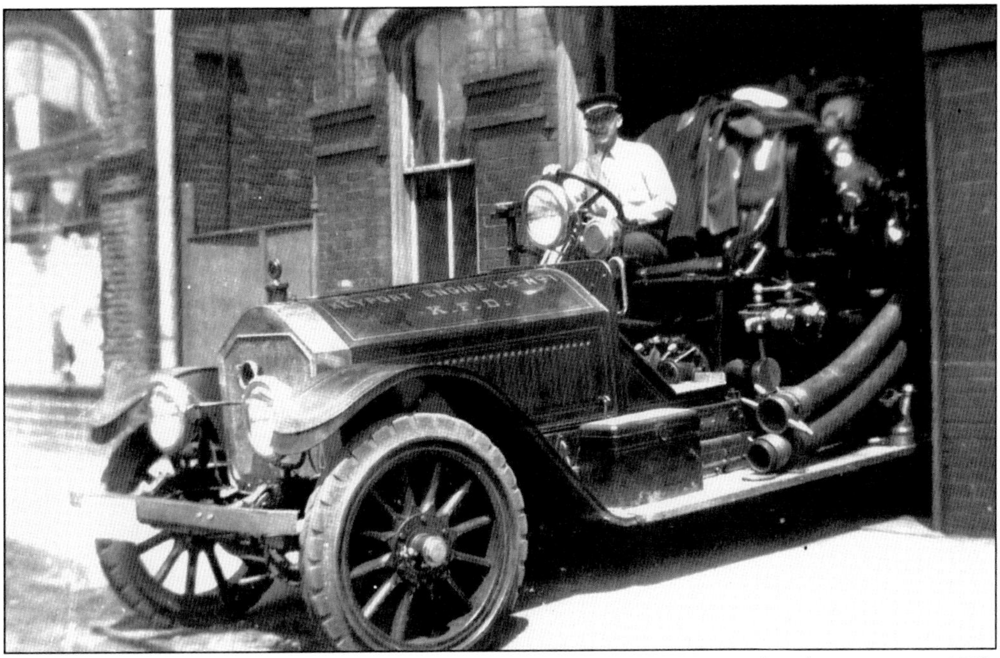

THE 1919 AMERICAN LAFRANCE. Keyport Engine Company men took a huge step in 1919, when they worked with the town to loan them half the money needed for Keyport's first pumping apparatus. This powerful 750-gallon-per-minute pumper was also used as the town's first ambulance. After proudly serving for nearly a quarter century, the American LaFrance was sent to the Aero Marine complex where its bronze pump and other items were scavenged for World War II scrap drives. Below, Bob Rose and Tony Diodato "respond" on the engine in June 1955.

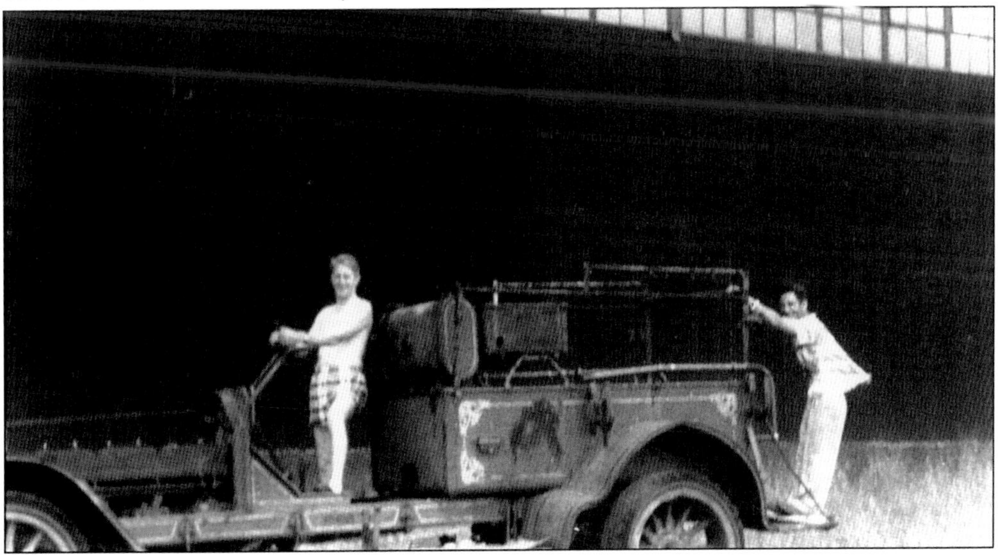

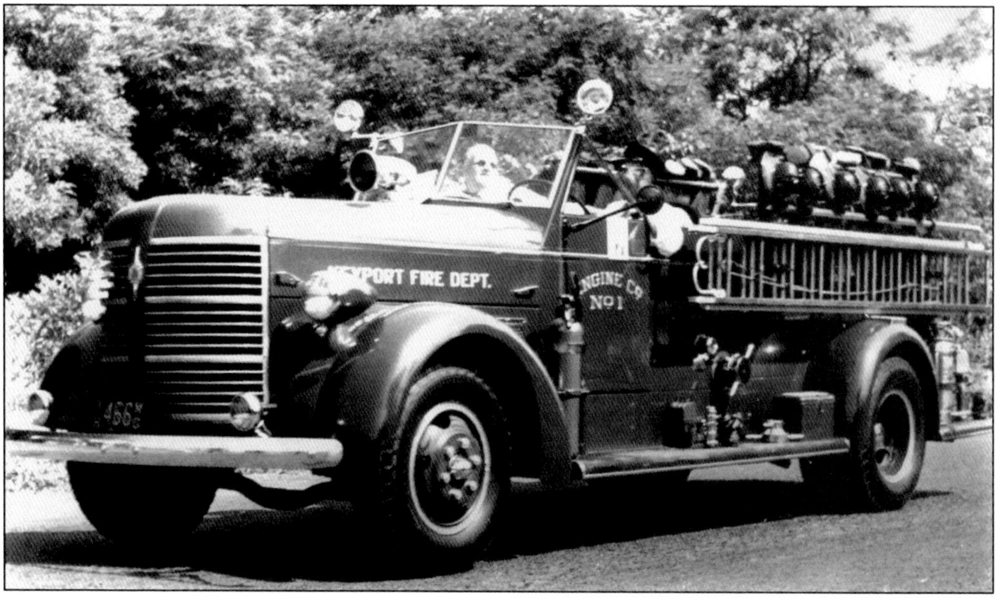

THE 1941 AMERICAN LAFRANCE. This classic American LaFrance 600-series pumper replaced the 1919 engine in 1941. Notice that the ladders are carried in the spot where the hard suction sleeve is normally located. American LaFrance would often move the pump panel to the officer's (right) side of the engine at the customer's request. The first two-way radio was installed in this rig in the early 1950s to be in constant communication with police dispatch.

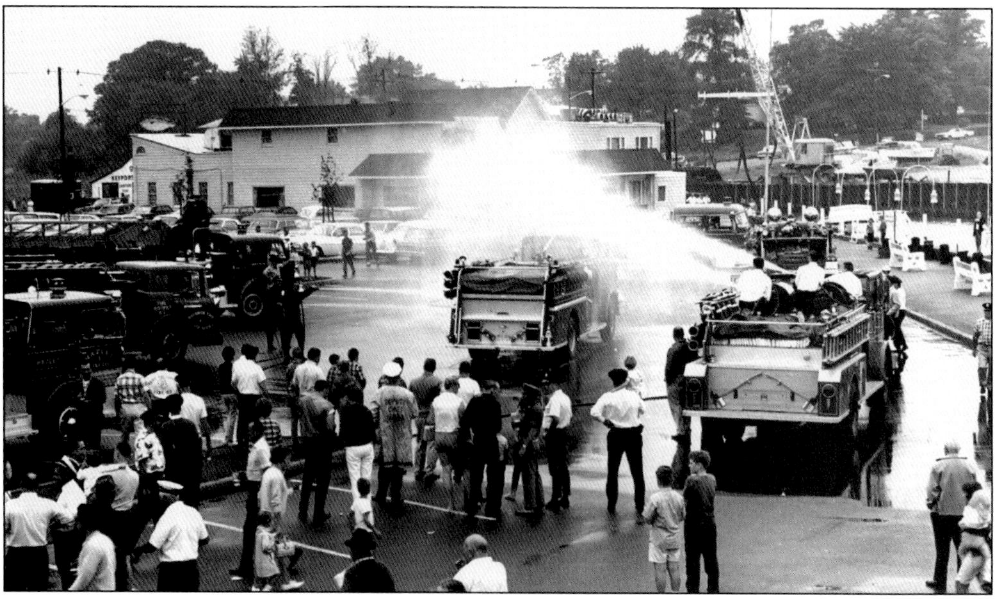

BRUCO WET DOWN, 1967. Each fire company in town works within a budget set forth by the governing body when purchasing a new apparatus. Specifications are developed over many months, and sometimes even years, to ensure that the new rig will be a good fit for the town and the company. Sealed bids are then advertised and opened, and a contract is awarded to a particular builder. Within 15 to 24 months, the new apparatus is delivered and welcomed to its new duty by a traditional wet down ceremony. (Courtesy Tom Gallo.)

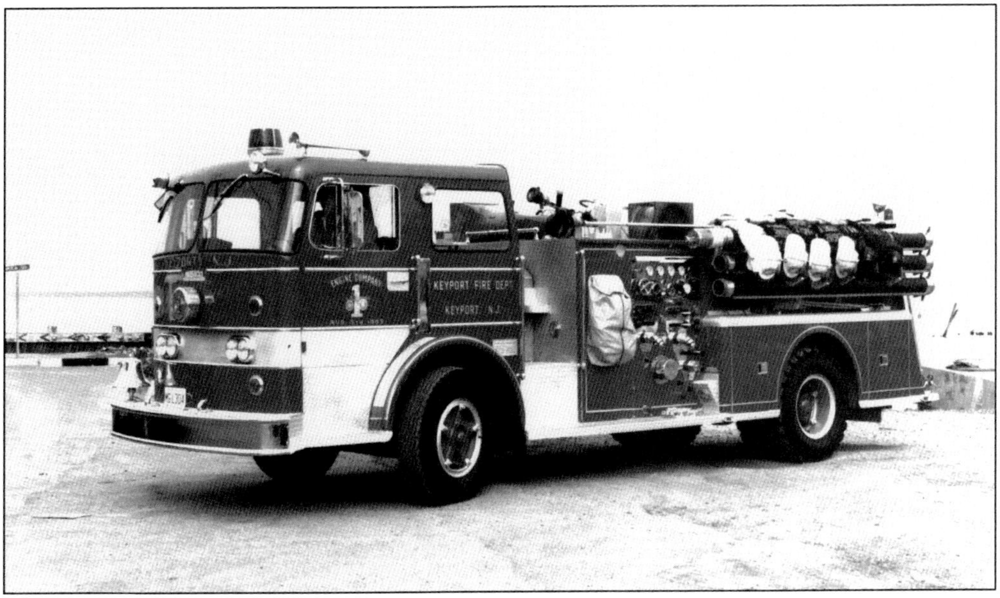

THE 1967 DUPLE/BRUCO. To this day, no apparatus was more customized than this rig of the Keyport Engine Company. Built on a Warner-Swasey heavy-duty chassis, with a duplex Cincinnati-style cab, the chassis was then sent to Altoona, Pennsylvania. Here the Brumbaugh Body Company completed the engine with a 1,000-gallon-per-minute, high-pressure pump as well as a Rockwood High Expansion foam system, which was one of the first in New Jersey. This engine has the distinction of answering more fire calls than all others, as it responded to all daytime still alarms for over 23 years.

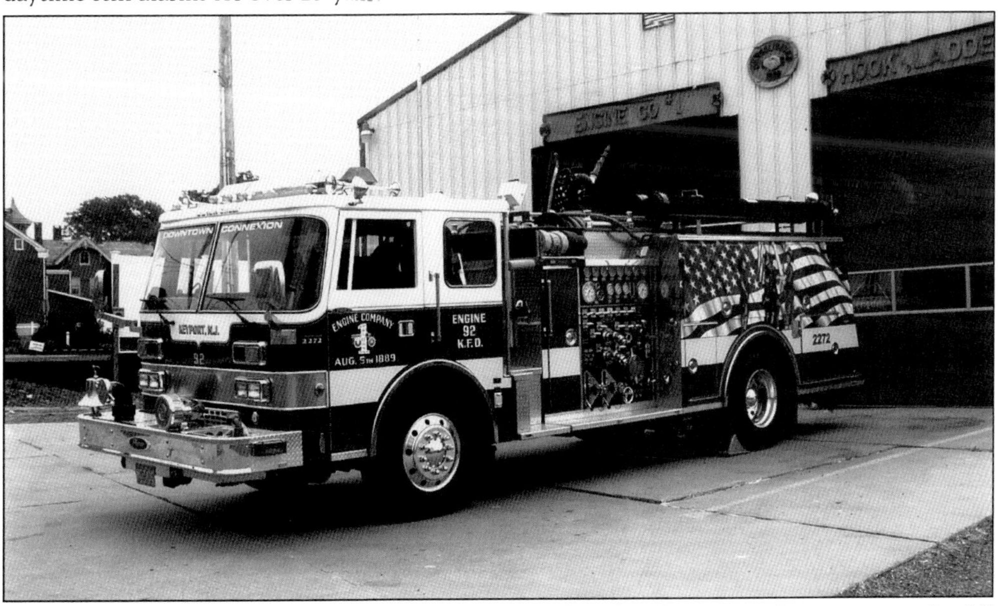

THE 1990 PIERCE. Keyport Engine Company received its fourth motorized apparatus in the fall of 1990. This engine was built by Pierce manufacturing in Appleton, Wisconsin, and cost just under $160,000. It was equipped with a 1,250-gallon-per-minute, three-stage, high-pressure pump and was among the last open jump seat apparatus delivered. Today all fire apparatus must have completely enclosed cabs, and no firefighter may stand or ride in an open position.

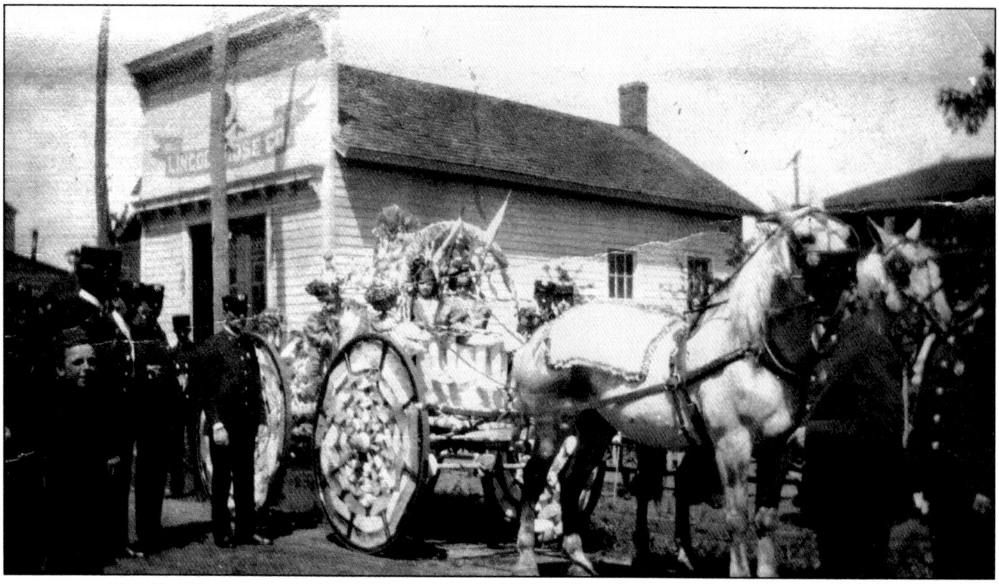

LINCOLN HOSE, 1895. Here the dapper whites (horses) of Lincoln Hose Company are hitched up to "Old Abe" for the big parade. Lincoln Hose Company affectionately named its first hose carriage after the company namesake and great American president when it was delivered in 1894. All three hose companies received identical four-wheel carriages that were built by the Keyport firm of Tilton and Cherry. In the background is a rare glimpse at the first hose house, which was built in 1894 on Waverly Street.

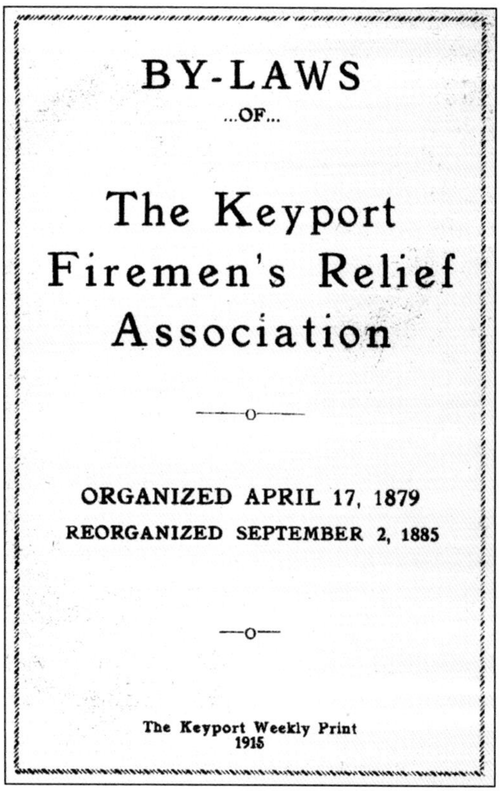

BY-LAWS

...OF...

The Keyport Firemen's Relief Association

——o——

ORGANIZED APRIL 17, 1879

REORGANIZED SEPTEMBER 2, 1885

—o—

The Keyport Weekly Print
1915

RELIEF BYLAWS, 1915. The Keyport Firemen's Relief Association was chartered in 1879 under state law. The Keyport Firemen's Relief Fund is administered both locally and through a state organization that provides death benefits for any member who has completed seven years of active service. Today relief benefits have grown from about \$50 in 1930 to almost \$10,000 in 2008.

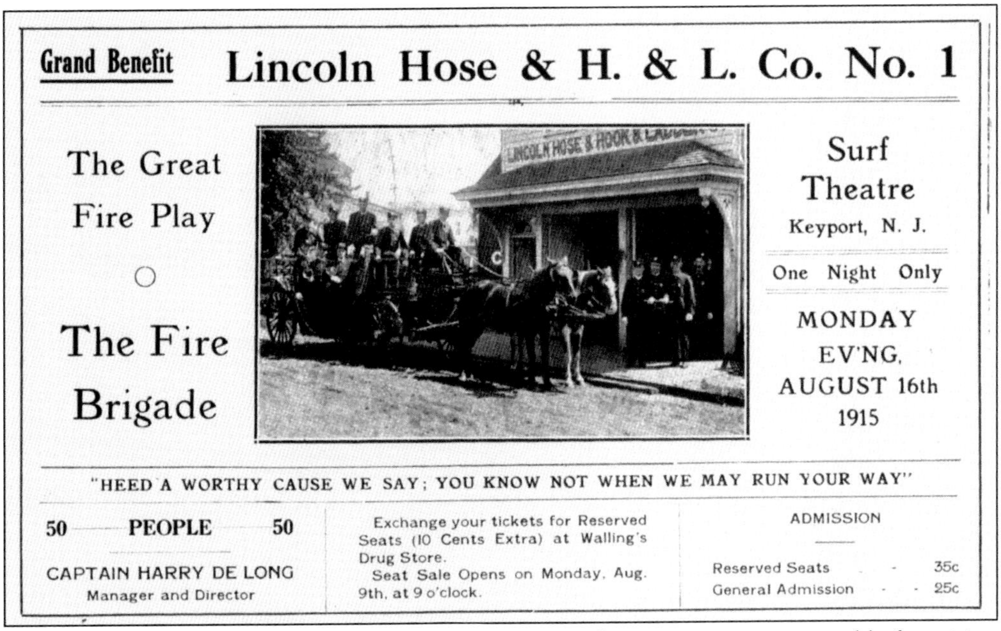

Grand Benefit Lincoln Hose & H. & L. Co. No. 1

The Great Fire Play

○

The Fire Brigade

Surf Theatre

Keyport, N. J.

One Night Only

MONDAY EV'NG, AUGUST 16th 1915

"HEED A WORTHY CAUSE WE SAY; YOU KNOW NOT WHEN WE MAY RUN YOUR WAY"

50 PEOPLE 50

CAPTAIN HARRY DE LONG
Manager and Director

Exchange your tickets for Reserved Seats (10 Cents Extra) at Walling's Drug Store.
Seat Sale Opens on Monday, Aug. 9th, at 9 o'clock.

ADMISSION

Reserved Seats - 35c
General Admission - - 25c

AUTO TRUCK FUND-RAISER, 1915. For many years, each company was responsible for raising funds for apparatus replacement. Lincoln ran many summer fairs as well as this grand vaudeville act in 1915 to generate income for a newfangled motor truck. The show must have done well, because three months later, Lincoln had purchased and rebuilt a 1912 Acme truck into a fine hose wagon. Although no photograph of this truck exists today, its arrival marked a new dawn for fire apparatus in Keyport.

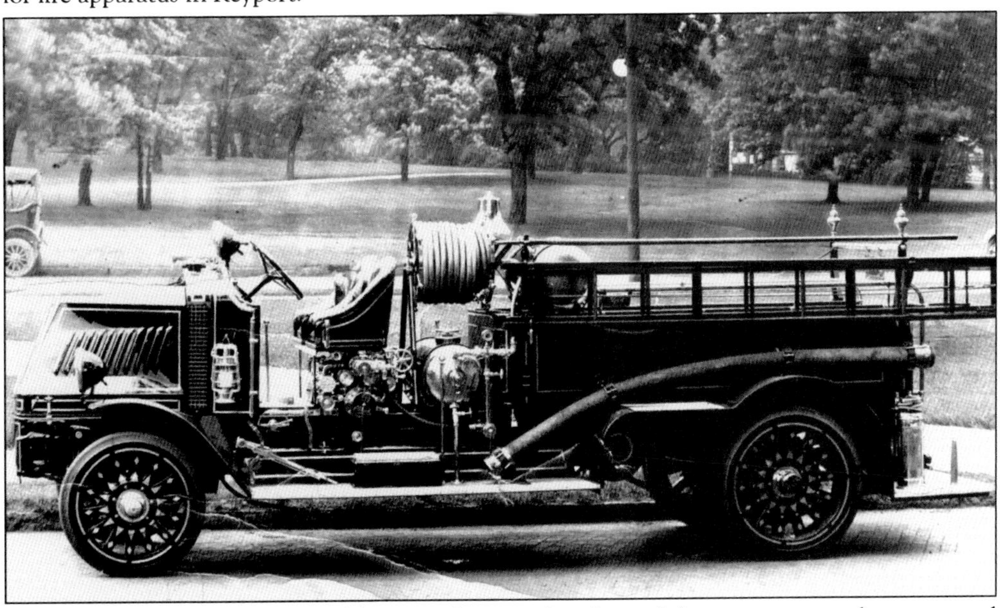

THE "BIG MACK." Lincoln had become frustrated with used fire apparatus and soon vowed to get a real fire engine. In 1921, this Mack AC 500-gallon-per-minute pumper was purchased with company funds. This massive pumper dwarfed nearly all vehicles on the road at the time, and it was quickly dubbed the "Big Mack." Operated for many years by Chief Engineer Charlie Henwood and his trusty dog Pal, the truck was sold as scrap for $500 in 1942.

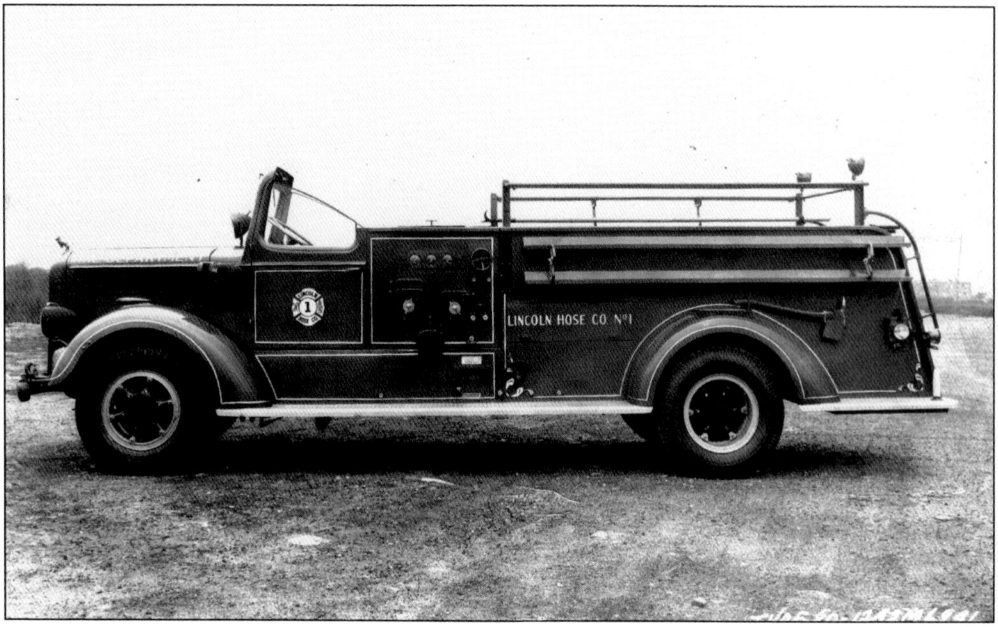

LINCOLN 1941 MACK. Lincoln Hose Company took delivery this of Mack LS-1129 model pumper in 1941. Equipped with a 750-gallon-per-minute pump, this engine replaced the 1921 Mack. Its streamlined shape was muted by the absence of any chrome or bright work due to wartime restrictions. This engine served with pride and was operated by Chief Engineer Allyn Lee for the 27 years it served Keyport. In 1969, the "'41 Mack" was replaced by another Mack pumper; below, it sits behind the firehouse, waiting to be disposed of in 1969. The 1941 Mack continued to serve Harris Gardens Fire Company in Union Beach for several years and was later sold to a private collector. Its whereabouts are unknown today.

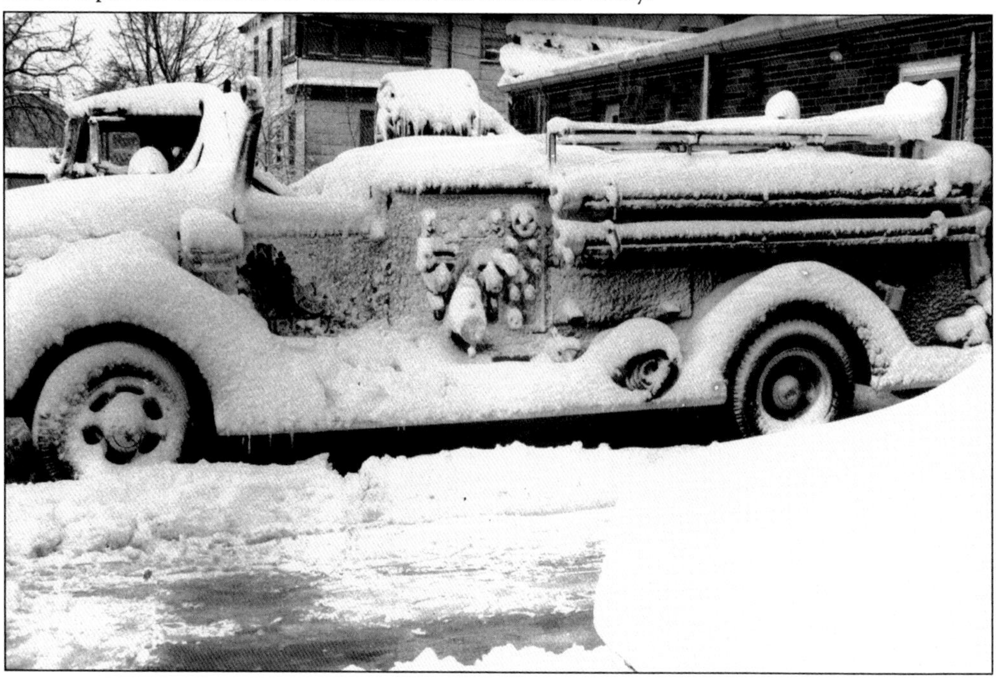

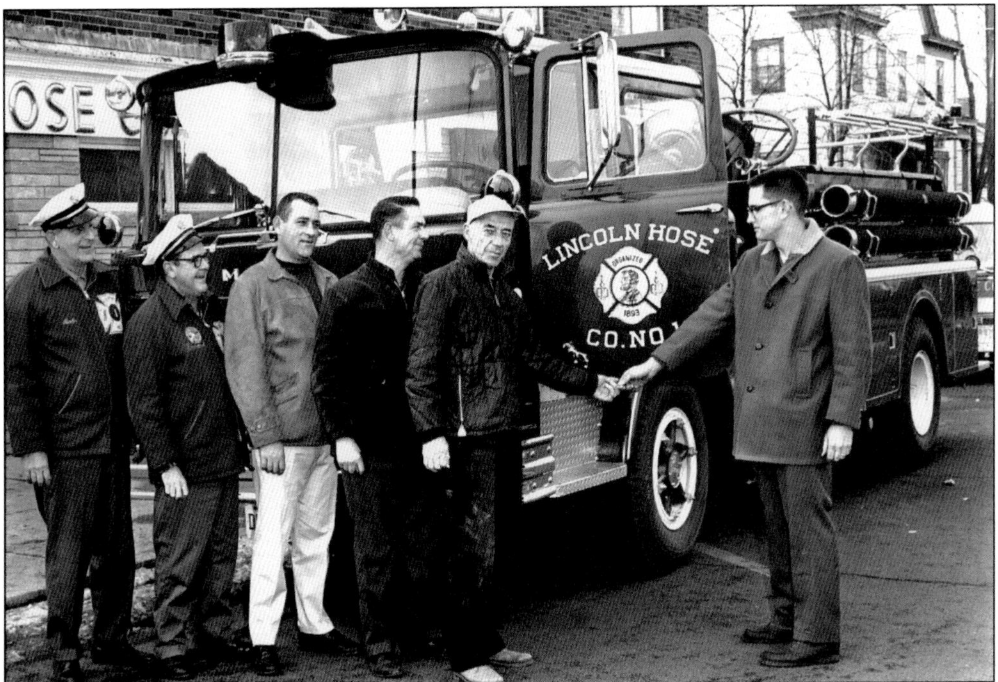

THE MACK CF. Although considered a 1968 model, the CF (cab forward) Mack did not arrive in town until the winter of 1969. Here, from left to right, Assistant Chief Bob Rose, Chief Howard Creed, ex-chief and engine designer Frank Van Pelt, Fire Commissioner Leroy Hicks, and Chief Engineer Allyn Lee accept the keys from Bill Phillips of Glasofer Mack in front of the Second Street firehouse.

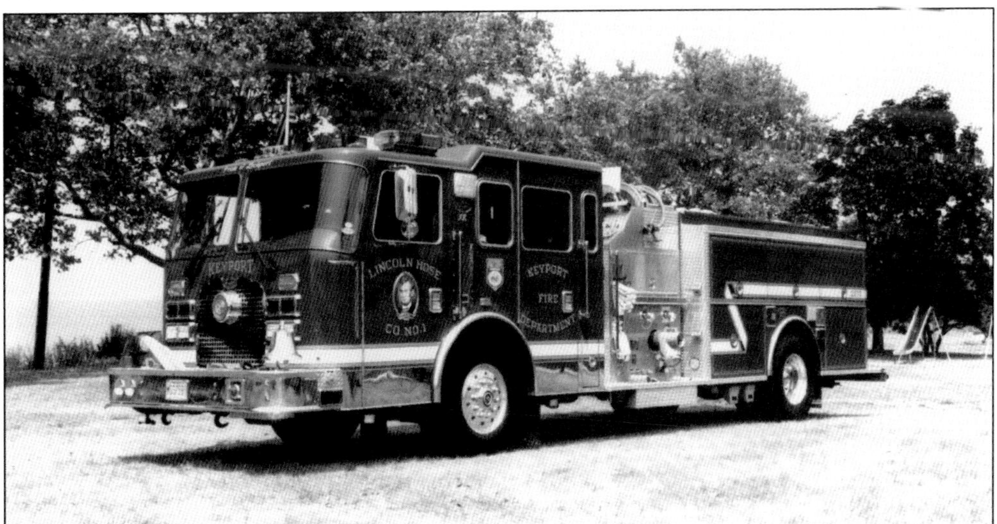

THE 1999 KME. It took over five years of hard work to develop specifications for a new engine for Lincoln. In 1999, the $276,000 contract was awarded to the Pennsylvania firm of Kovatch Mobile Equipment (KME) to build this engine. This apparatus is equipped with a 1,500-gallon-per-minute pump and holds nearly 1,000 feet of large-diameter hose and carries 10 firefighters in the large cab. This engine was dedicated to Frank "Junior" Van Pelt, ex-chief and life member, who was the chairman of the truck committee.

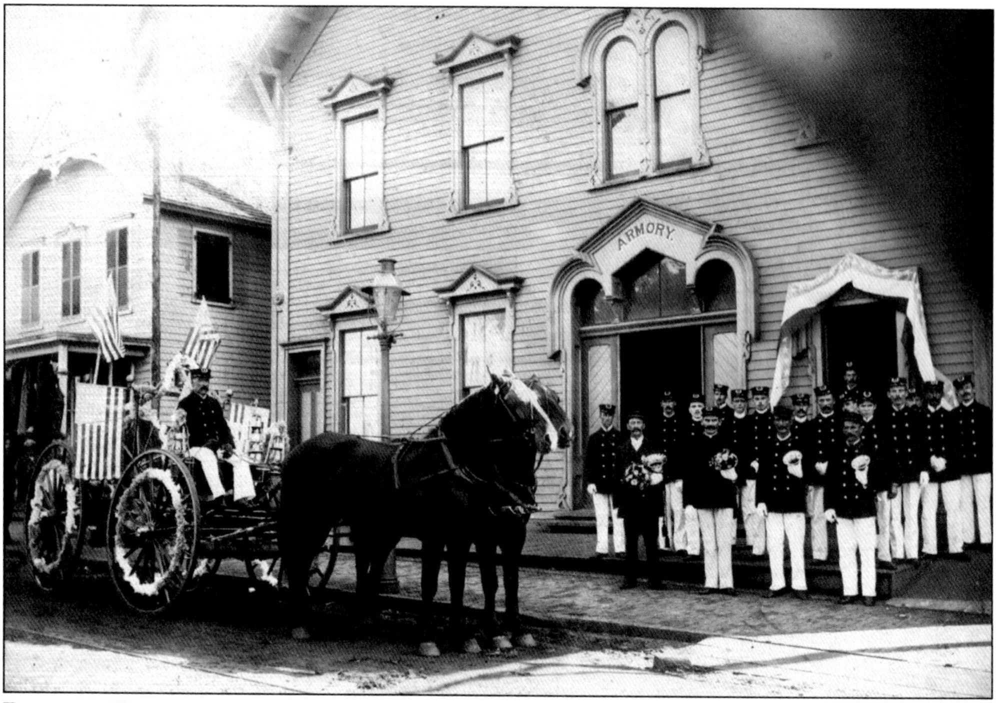

RARITAN HOSE IN 1895. Raritan Hose Company began on May 29, 1893, when many of these men gathered in the armory hall to organize a hose company in the business district. Their first uniforms consisted of sweaters with "R. H." embroidered on them but soon switched to this attire of white pants and a double-breasted coat. At the reins of the carriage is Capt. Oscar F. Stanhope; today the hat he is wearing in this photograph is on display in the KFMA.

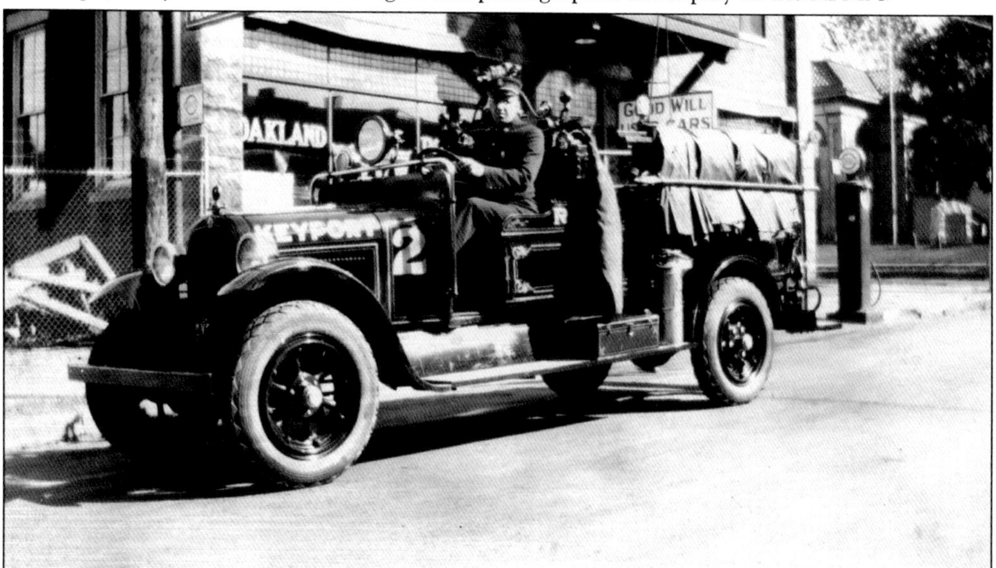

RARITAN'S GRAHAM HOSE WAGON. Taken on Broad Street directly across from the company quarters is Raritan Hose Company's second motor truck. After a few years of running a used Rainer hose wagon (of which no known photograph exists), the company purchased this rig in 1928. It would serve as Keyport's first floodlight unit in later years and was replaced in 1946.

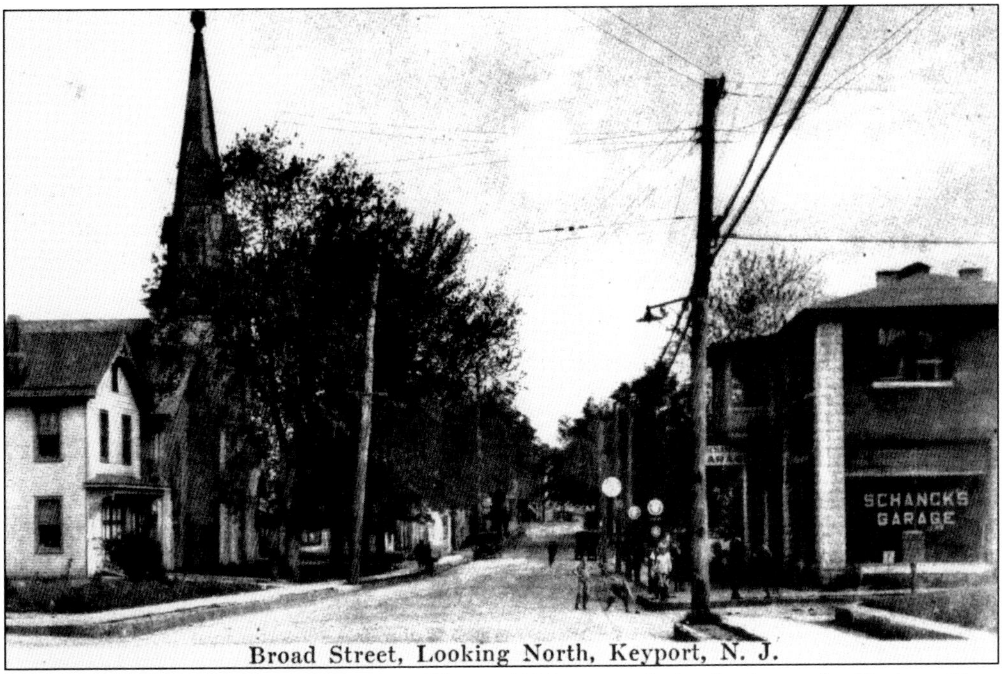

RARITAN HOSE AND BROAD STREET. The *c.* 1920 postcard clearly shows the Raritan Hose Company firehouse on the left and the general neighborhood heading uptown. This firehouse was built in 1900 on Division Street and moved to this location in 1919. It would continue to serve as Raritan Hose Company's quarters until 1969. In 2002, it was obtained through a state grant and became the KFMA. (Courtesy Tom Gallo.)

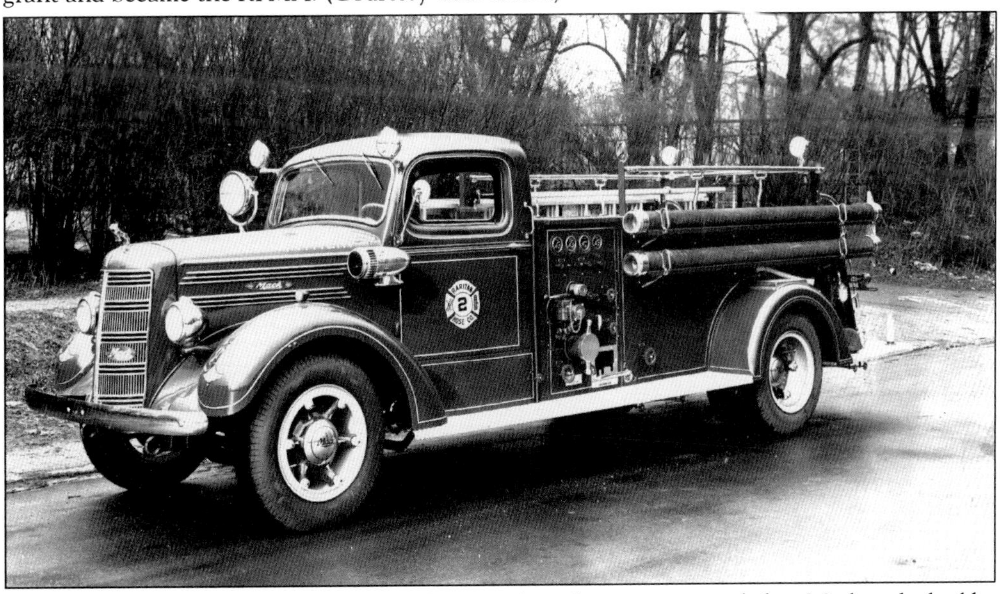

THE 1946 MACK PUMPER. Raritan Hose Company's truck committee settled on Mack as the builder of its first pumping fire apparatus. This smaller size E model contained a 500-gallon-per-minute centrifugal pump as well as a small booster tank. Note the scrollwork on the fenders, which was an ingenious way to dress up wartime fire apparatus that were devoid of chrome. This engine served until 1974 and remained in Keyport until 1989, when it left for Vermont on a flatbed truck.

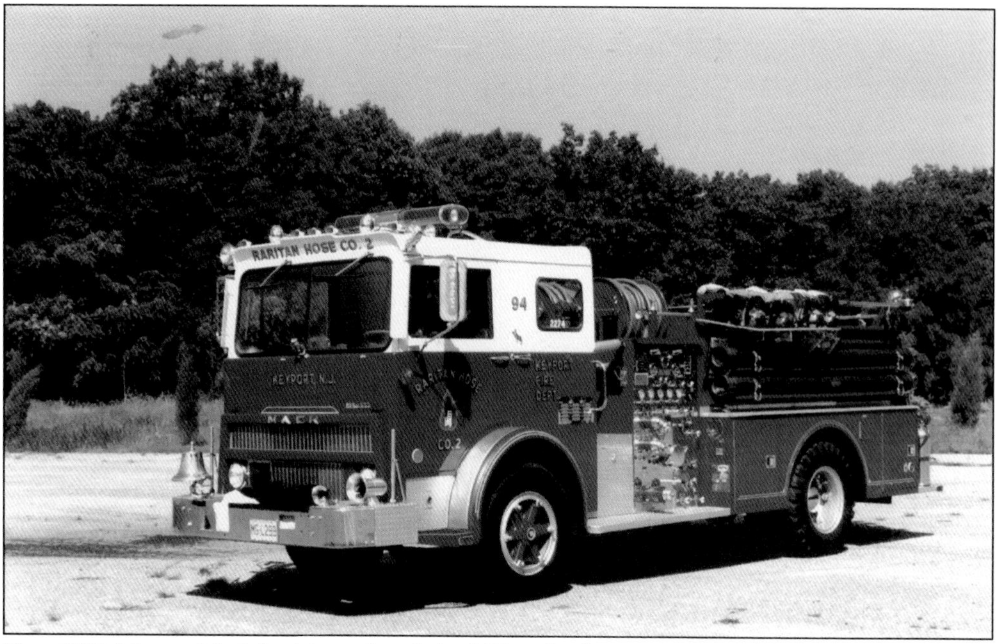

The 1974 Mack. Mack had introduced the MB-style chassis in 1966, although it did not gain popularity as a fire apparatus until the mid-1970s. The Borough of Keyport purchased this Mack MB 1,000-gallon-per-minute pumper with bodywork by Howe in 1974. It was the first engine in town to be painted red and white, as well as the first to be equipped with only an electronic siren. Designated "74," it served faithfully for 29 years.

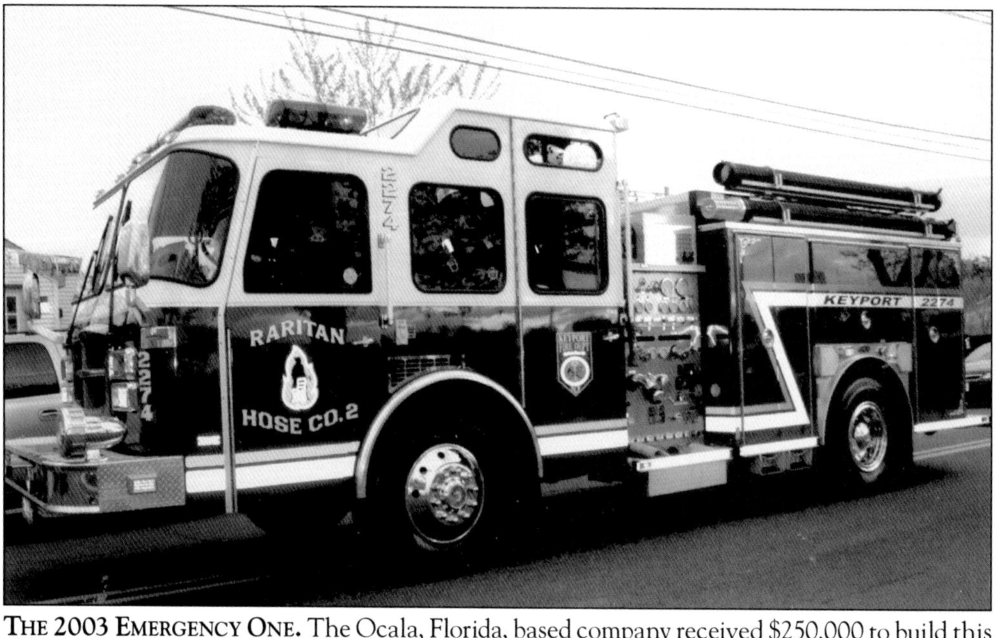

The 2003 Emergency One. The Ocala, Florida, based company received $250,000 to build this good-looking engine in 2003. Fitted with a 1,250-gallon-per-minute pump as well as 1,000-gallon water tank, this engine is technically classified as a pumper-tanker. Engine 74 is also equipped with a Compressed Air Foam System (CAFS), which has proven to have a quicker knockdown time on fires by mixing foam, water, and air as the firefighting agent.

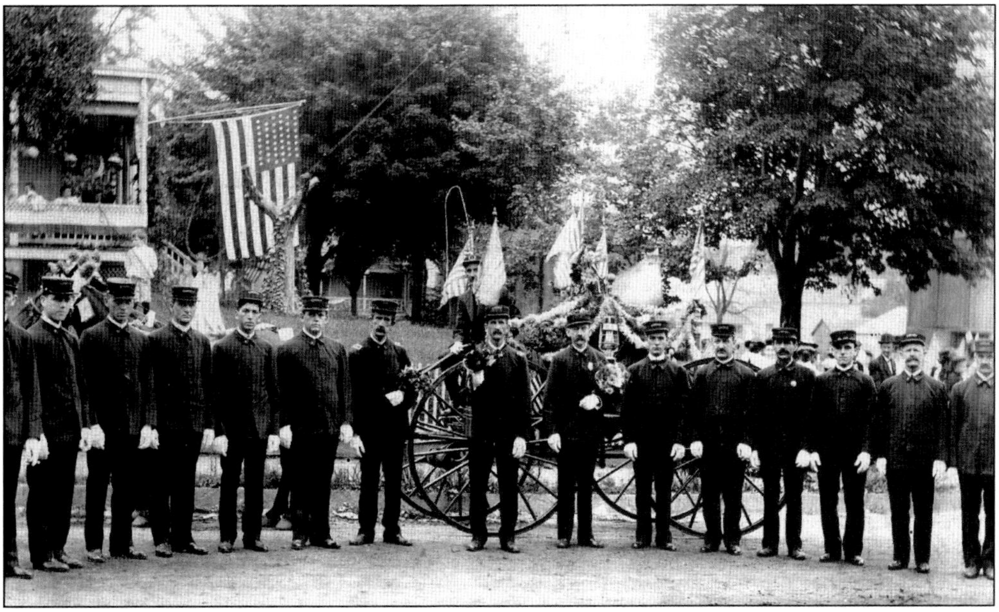

Liberty Carriage, 1894. This is the earliest known view of Liberty Hose Company No. 3, which began service on June 28, 1893, in the Mechanicsville section of town. By July 4 weekend of that year, the company, under the guiding hand of member and contractor Gordon Davison, had erected a fine one-story hose house on Atlantic Street. Soon after, they took delivery of this four-wheel service hose carriage, which was built by in Keyport by Tilton and Cherry.

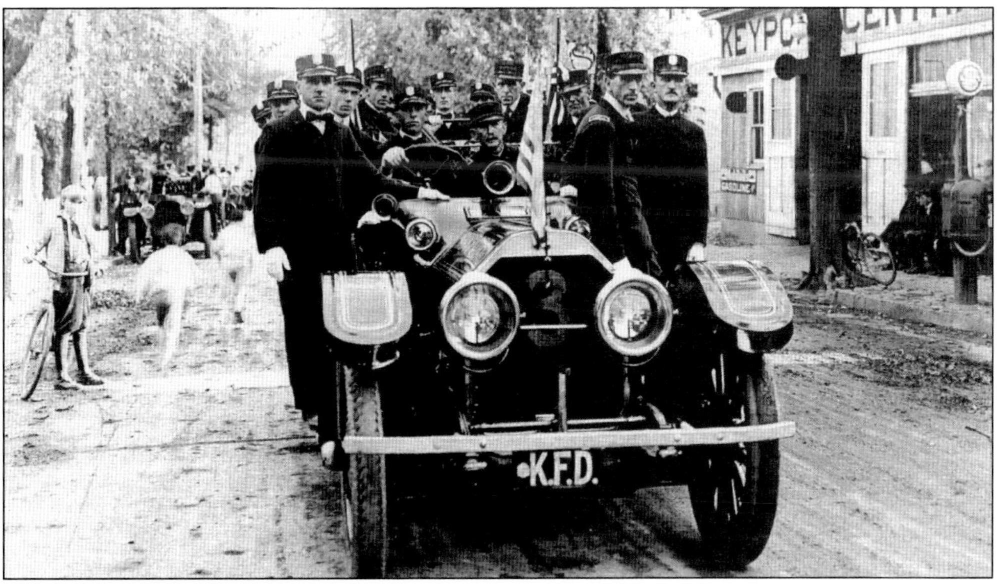

The 1913 Oldsmobile. In 1916, Liberty Hose Company purchased this used Oldsmobile touring car, which was promptly turned into a hose wagon by company member Alvarado Walling. In just a few years, this truck would hold the New Jersey State Hose Laying Speed record, which remained unbroken for many years. This little truck was extremely fast and often responded to fires in Holmdel and Colts Neck, long before those towns had motorized equipment. Liberty Hose Company still maintains a large collection of trophies and silver cups that were won by this hose team and the Oldsmobile.

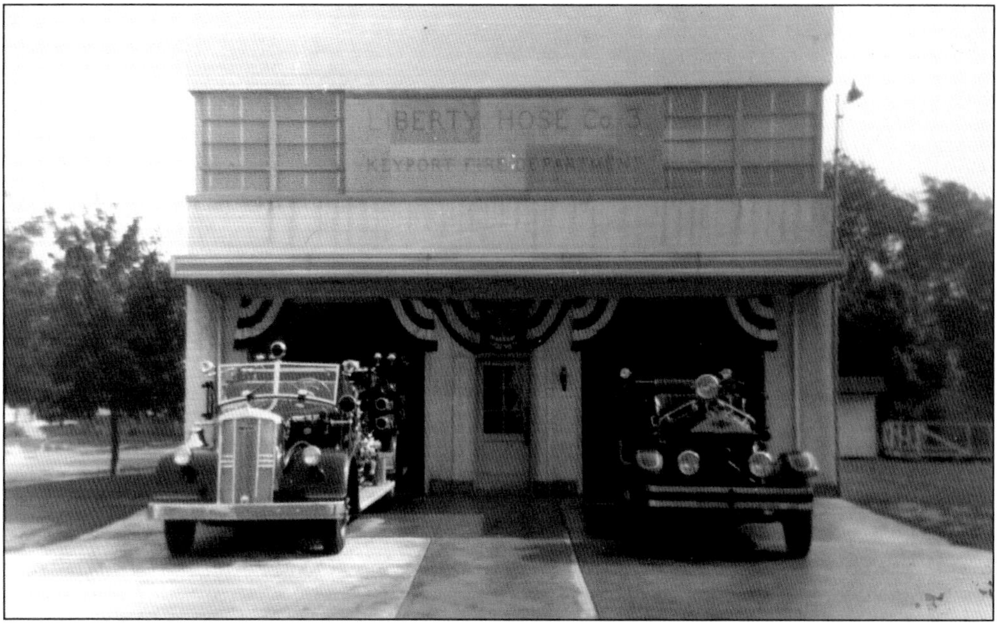

THE 1948 FIREHOUSE. By 1948, the company was feeling cramped by the old firehouse and set out to purchase a lot for a new building. Soon this building was completed at the corner of Atlantic Street and State Highway 36. It became the most modern firehouse to date and contained a large meeting hall, commercial kitchen, and two apparatus bays. Here the 1952 and the 1929 share a brief moment together. By 1996, these two engines would be reunited and sit side-by-side inside the new Liberty Hose Company firehouse, which was completed in 1987.

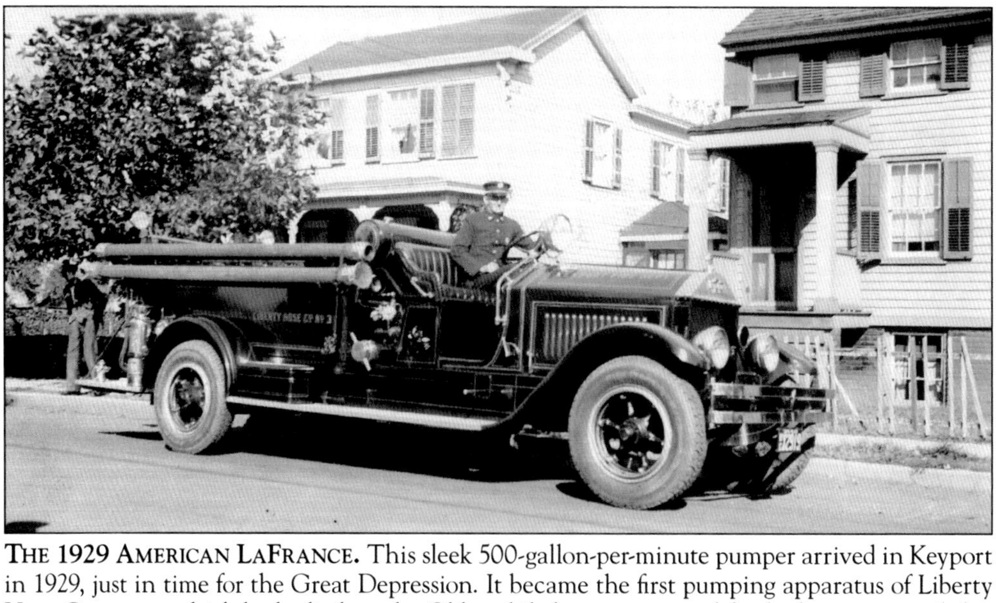

THE 1929 AMERICAN LAFRANCE. This sleek 500-gallon-per-minute pumper arrived in Keyport in 1929, just in time for the Great Depression. It became the first pumping apparatus of Liberty Hose Company, which had relied on the Oldsmobile hose wagon and fire hydrants up until this time for water pressure. The 1929 would spend its career in South Keyport until 1952, when it was sold to the Doris Duke estate in Somerset County as part of her private fire department. In 1996, Liberty Hose Company bought the truck from the Flagtown Fire Company and began a complete restoration of the truck, which was completed in 1998.

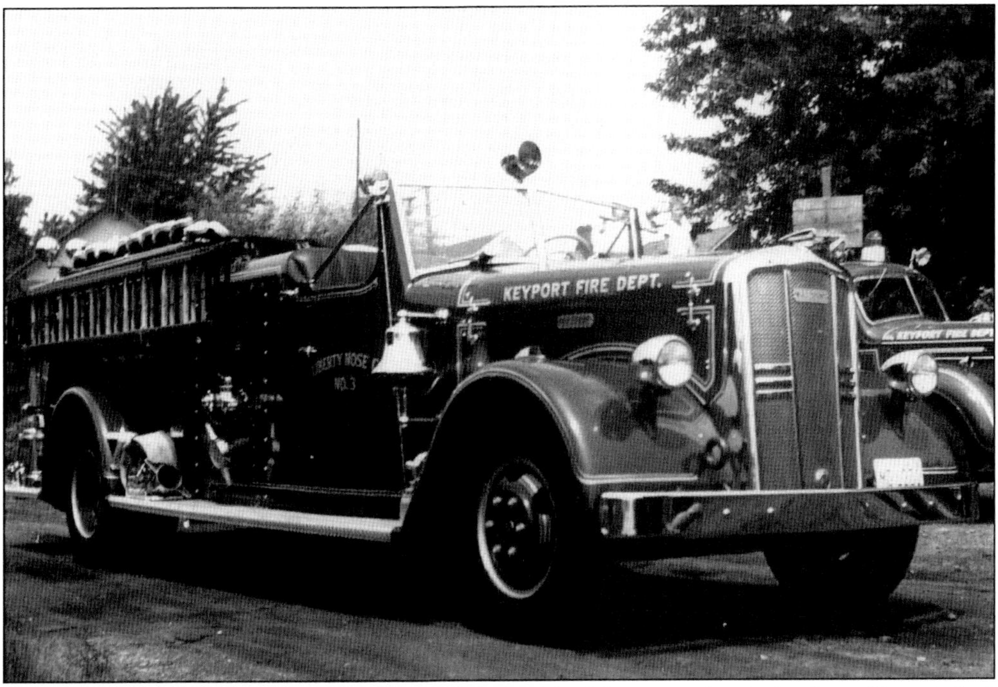

The Ward LaFrance. By 1952, the 1929 engine had been rendered obsolete. Since it had a completely open cab and small pump, it would have to be replaced for comfort and safety. So in 1952, the borough purchased Liberty Hose Company this Ward LaFrance Model NT 750-gallon-per-minute pumper, which was built in Elmira Heights, New York. It should be noted that American LaFrance and Ward LaFrance were completely separate companies located in adjacent towns.

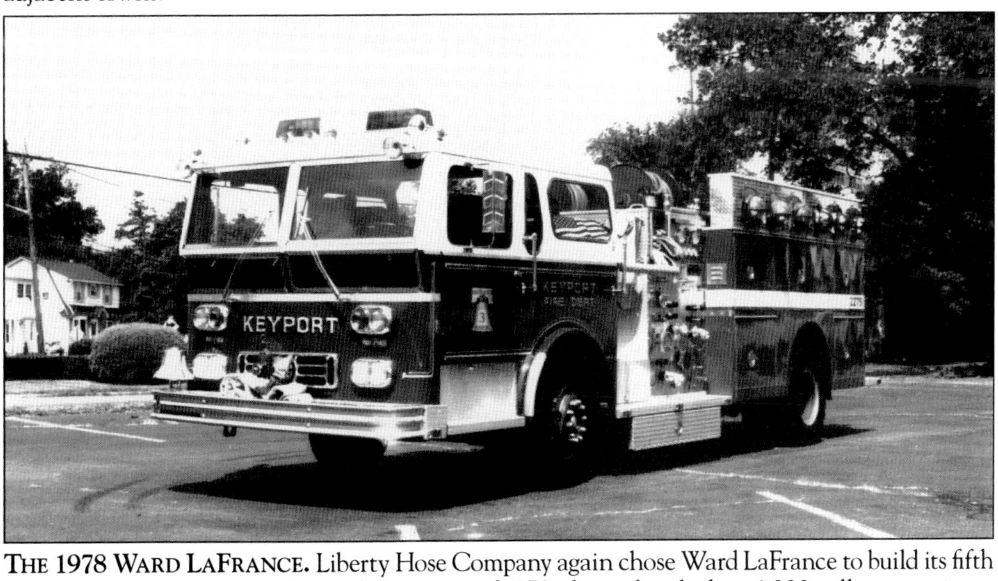

The 1978 Ward LaFrance. Liberty Hose Company again chose Ward LaFrance to build its fifth piece of fire apparatus in 1978. By the summer of 1978, this red and white 1,000-gallon-per-minute Ambassador Model engine was dedicated with a well-attended wet down on the firehouse grounds. It has been well maintained by the company and has been upgraded through the years to remain efficient. It is due to be replaced in the near future with a new apparatus.

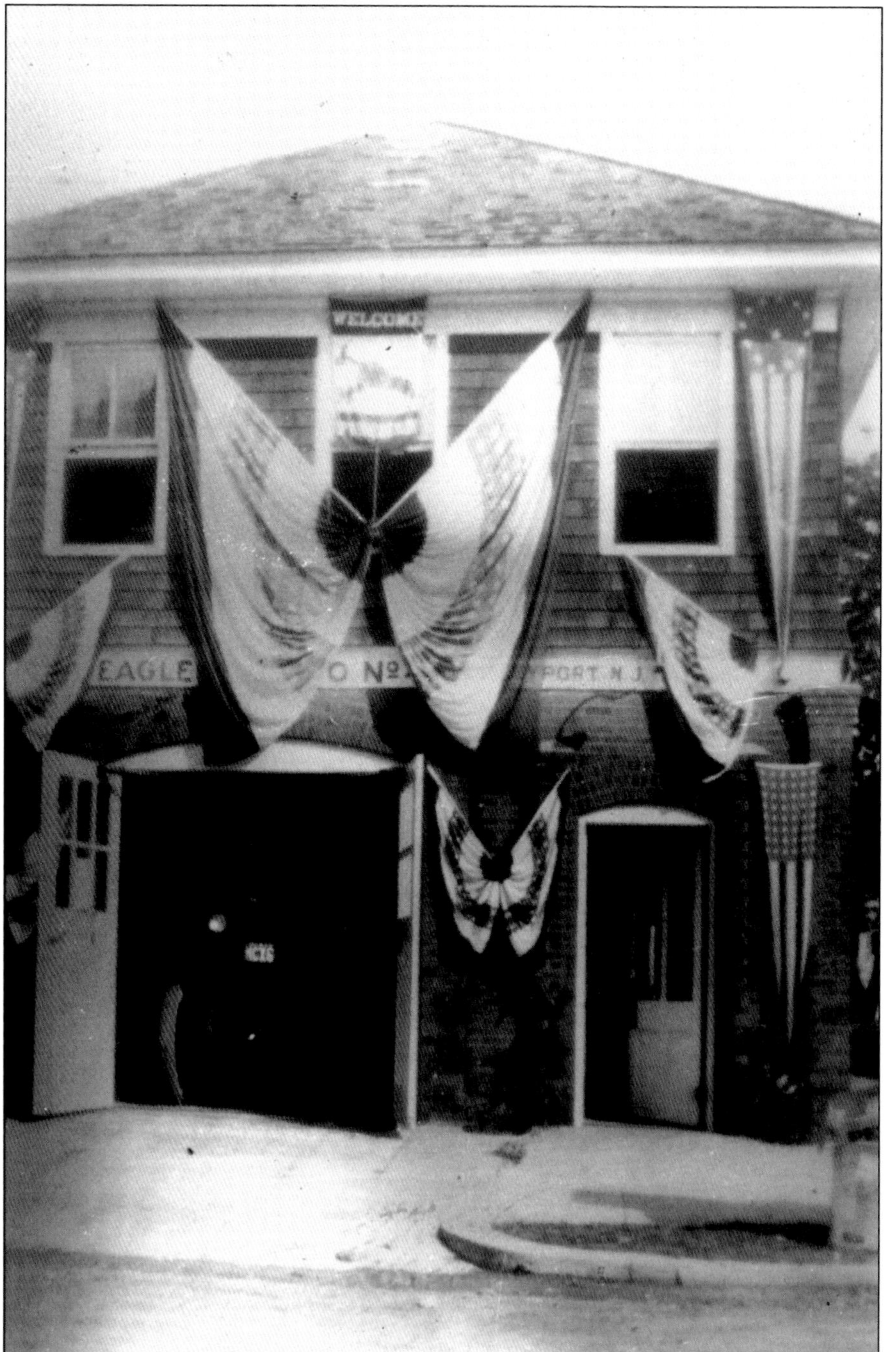

EAGLE HOSE COMPANY NO. 4. The western end known as the Brown's Point section of town was without fire protection until 1907. It was in that year that the Eagle Hose Company formed in the rooms of the Disbrow Bottling Company. It soon rented a barn on the Hopla property and in 1912 built this two-story hose house at 95 Broadway. For nearly 100 years, the company has endeavored to maintain this historic firehouse, and today it is one of the oldest active firehouses in Monmouth County.

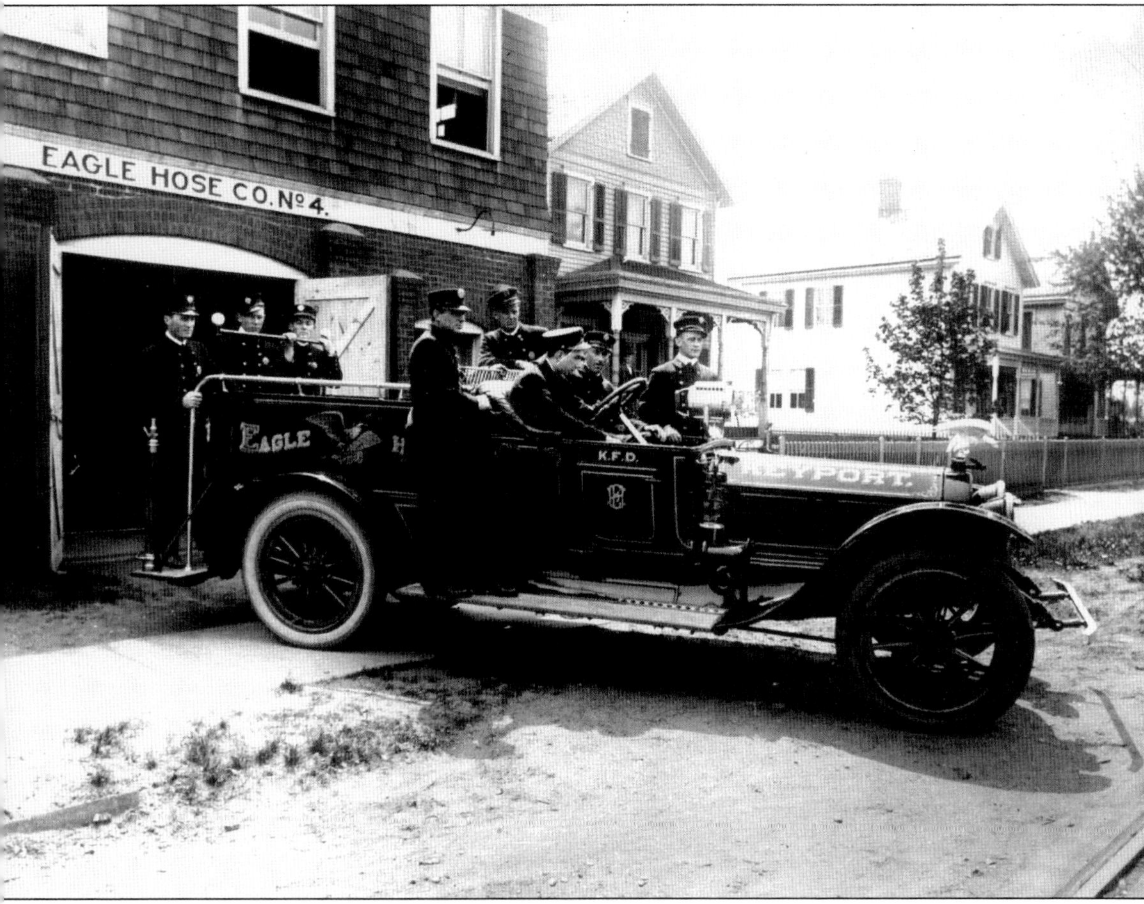

PEERLESS HOSE WAGON, 1916. The whole gang is aboard for this photograph, which was taken in front of the Broadway firehouse. This hose wagon started life as a luxury automobile and was later converted into this fancy yet practical firefighting apparatus by a local business. Although the Peerless is long gone the large, hand-carved wooden hood ornament is still on display in the company's quarters.

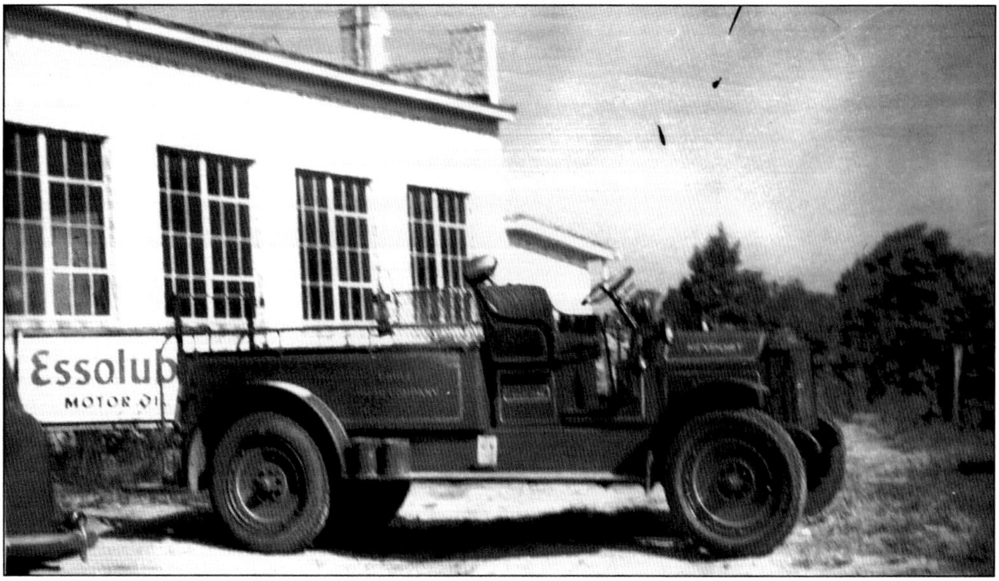

FEDERAL HOSE WAGON. In 1922, this Federal truck was procured to replace the 1915 Peerless hose wagon. The hose body was most likely transferred from the Peerless to make a functional yet simple fire apparatus. This truck served until 1937, when it was replaced by an International truck with Buffalo Fire Equipment bodywork. Here the old Federal sits outside the Raritan Garage, International truck dealer, after being traded in.

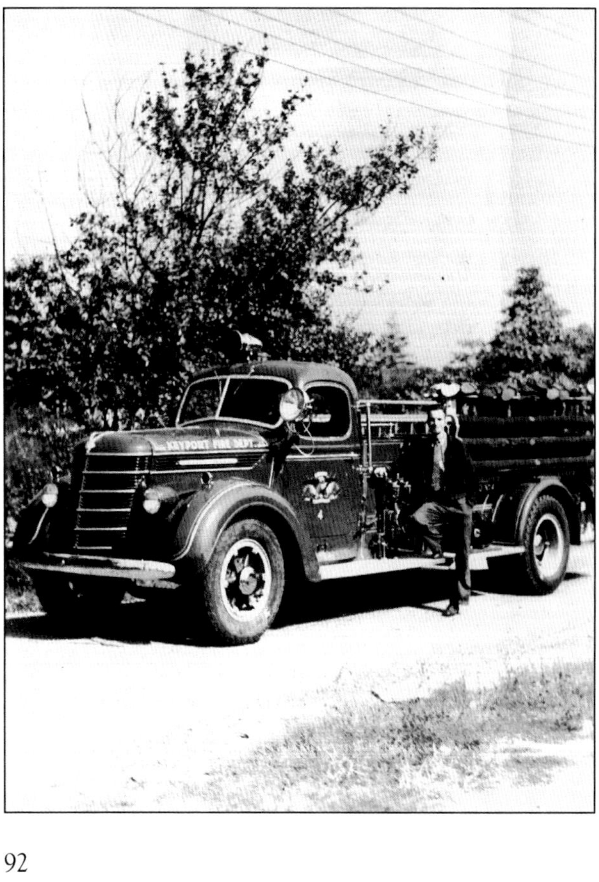

THE 1937 BUFFALO PUMPER. Eagle Hose Company took delivery of this streamlined Buffalo Fire Equipment pumper in 1937. Built on an International truck chassis, this rig was capable of pumping 500 gallons of water per minute. It was the first closed-cab pumper in the department and served the borough until 1962. (Photograph by James Lewis; courtesy KHS.)

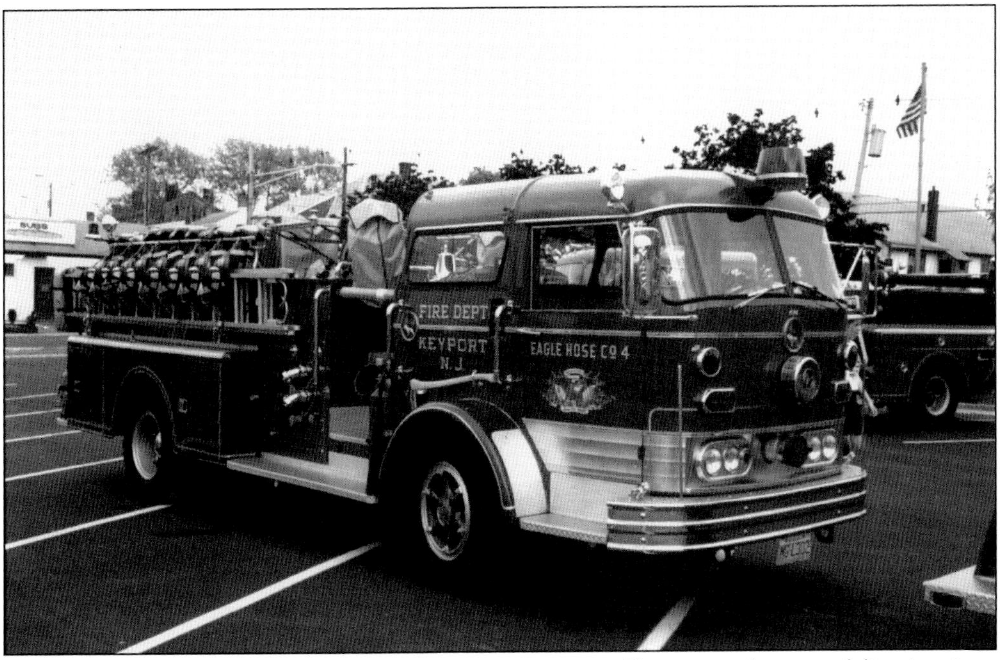

MACK C-MODEL. Mack fire apparatus was no stranger to Keyport with seven deliveries since 1921. In 1962, this C-Model, 1,000-gallon-per-minute pumper was delivered to the Broadway firehouse and placed in service soon after. For 27 years, truck 96 responded to fire calls in Keyport and was placed out of service in 1990.

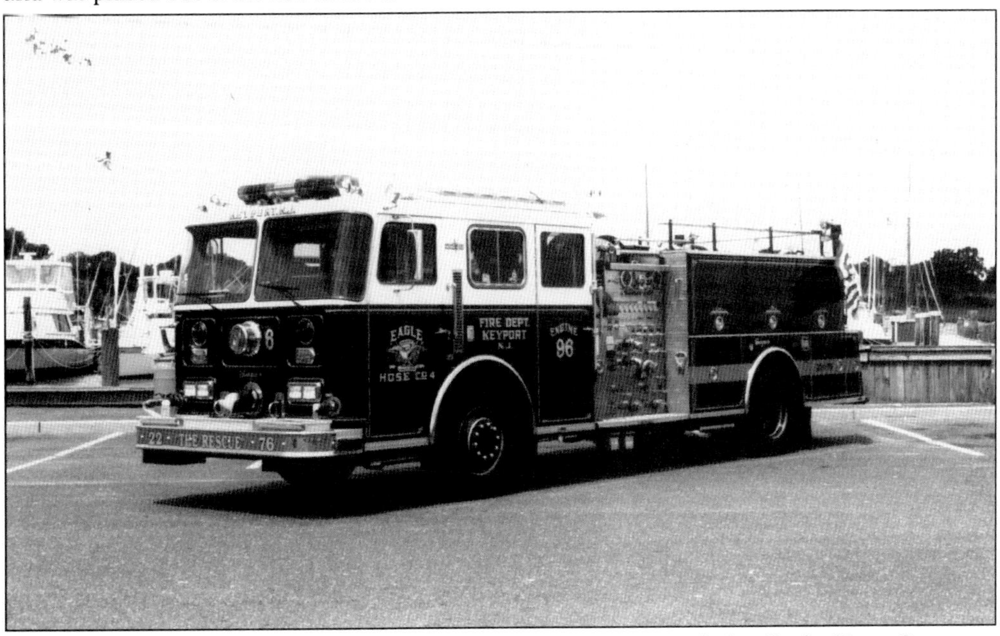

THE 1990 SEAGRAVE. Both the Keyport Engine Company and the Eagle Hose Company received new apparatus in 1990. The Borough of Keyport purchased this Seagrave fully enclosed pumper for just over $200,000 (about the same amount lost in the 1877 fire). It was built at the Seagrave factory in Clintonville, Wisconsin. In 1991, this apparatus was fitted with the Amkus Jaws of Life system and has responded to countless accidents along Keyport's busy highways.

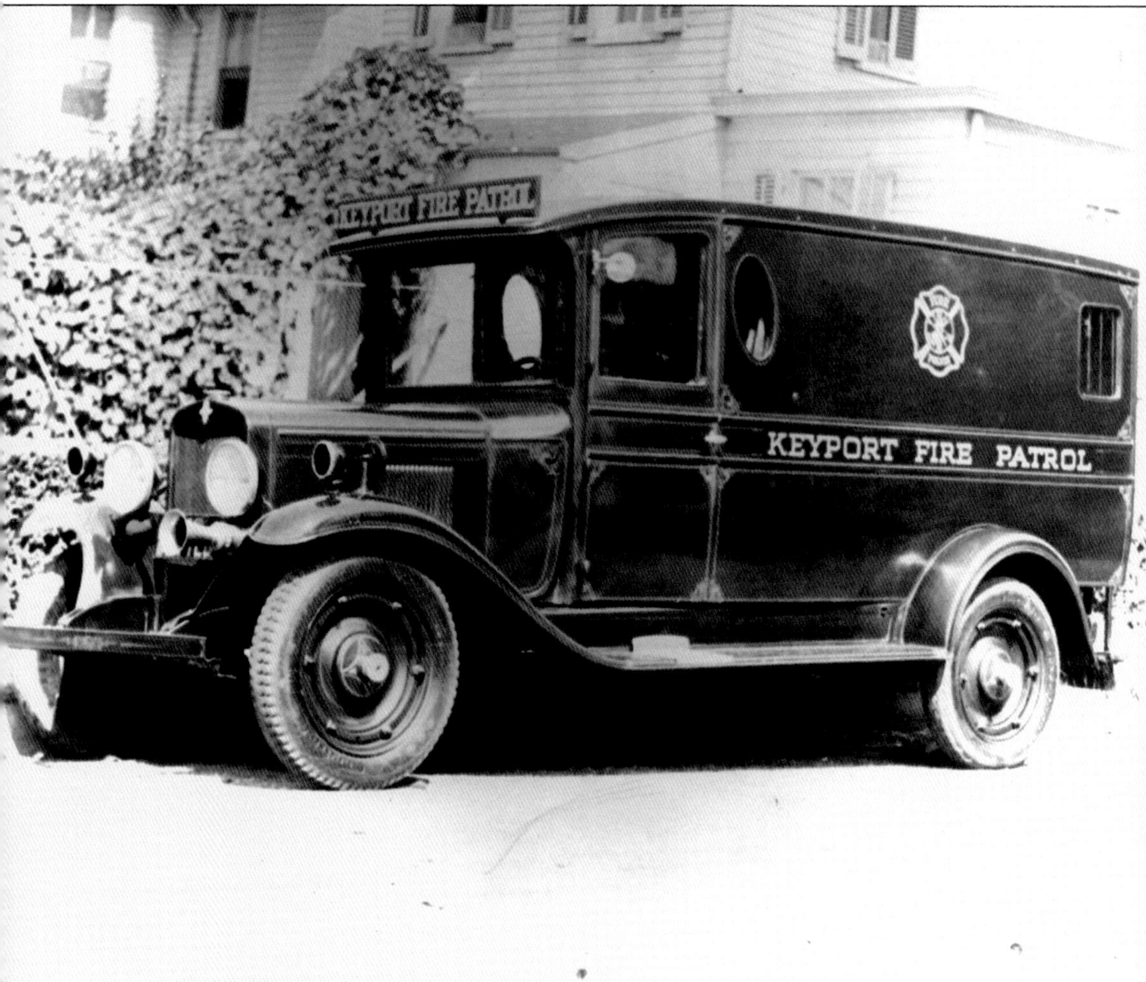

THE 1930 CHEVROLET. In 1930, this Chevy panel truck assumed the duties from the 1925 Ford. This old rig, which was referred to as the "Pie Wagon," would serve the borough until 1945. In its retirement years, it spent its days delivering ice around town operated by the Lawson family.

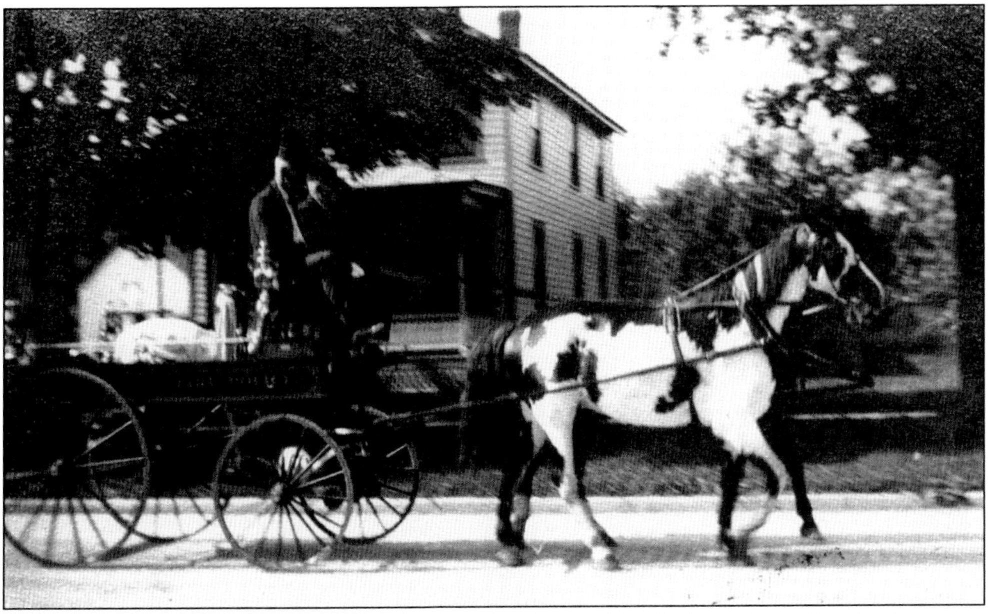

Fire Police Wagon, 1893. The Keyport Fire Patrol was officially formed in 1893. At that time, two members from each of the five companies volunteered to preserve peace, arrest violators, and salvage personal property. In 1912, the fire patrol, or fire police, became the seventh company of the department and was officially named the Board of Fire Wardens. In 2007, this name was officially changed back to Keyport Fire Patrol, and the term *fire warden* was relegated to the history books.

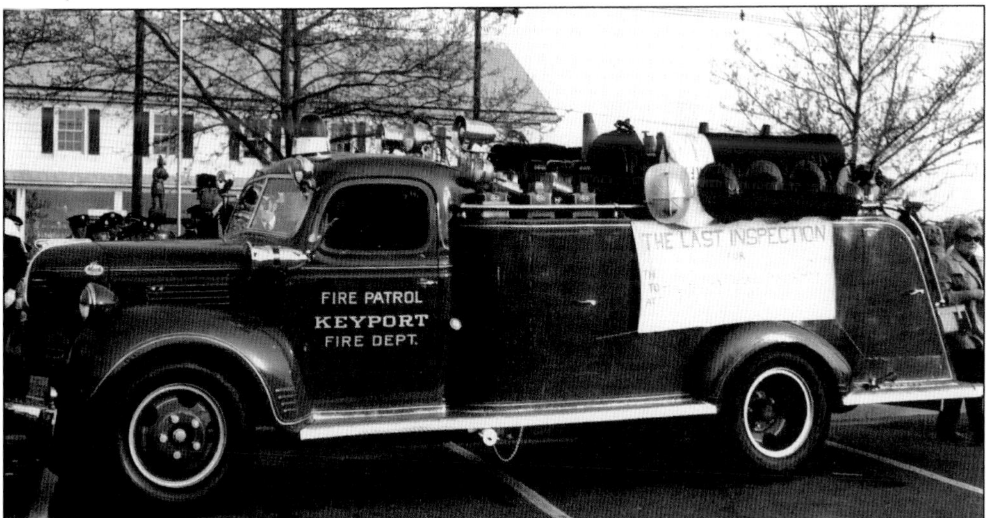

Dodge Fire Patrol Truck. In 1946, the borough purchased this custom-built Tasc pumper on a three-ton Dodge chassis. For purposes of civil defense, it was mandated that all fire apparatus be equipped with firefighting pumps in case of a nuclear attack. Since this time, the fire patrol has operated as an engine company in addition to its other specialized duties. In 1972, the Dodge made its last inspection showing; the sign on the side reads, "The last inspection for the little fire truck that does not pump to be replaced by a 1972 Hahn after 30 years of faithful service . . . Fire Patrol."

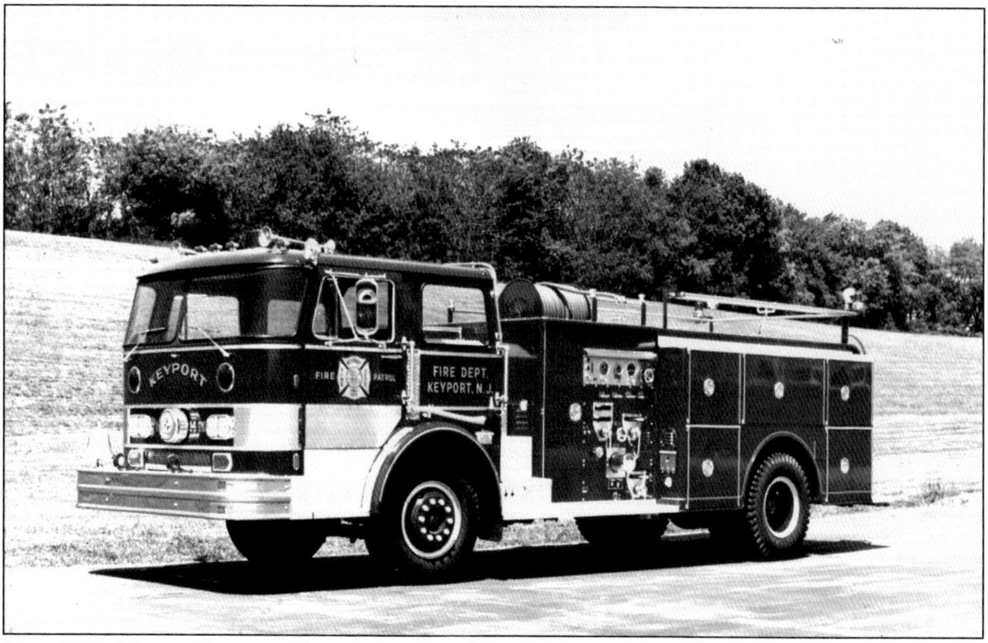

THE 1972 HAHN PUMPER TRUCK 97. Capable of pumping 750 gallons per minute, this truck contained a mobile air cascade system, which was one of the first in the Bayshore area. It also had two booster reels, one mounted on top and the other recessed above the rear step. After many years of busy work, the gasoline engine gave out, and the truck was deemed not to be repairable. It was sadly stripped of all usable parts and scrapped in 2002.

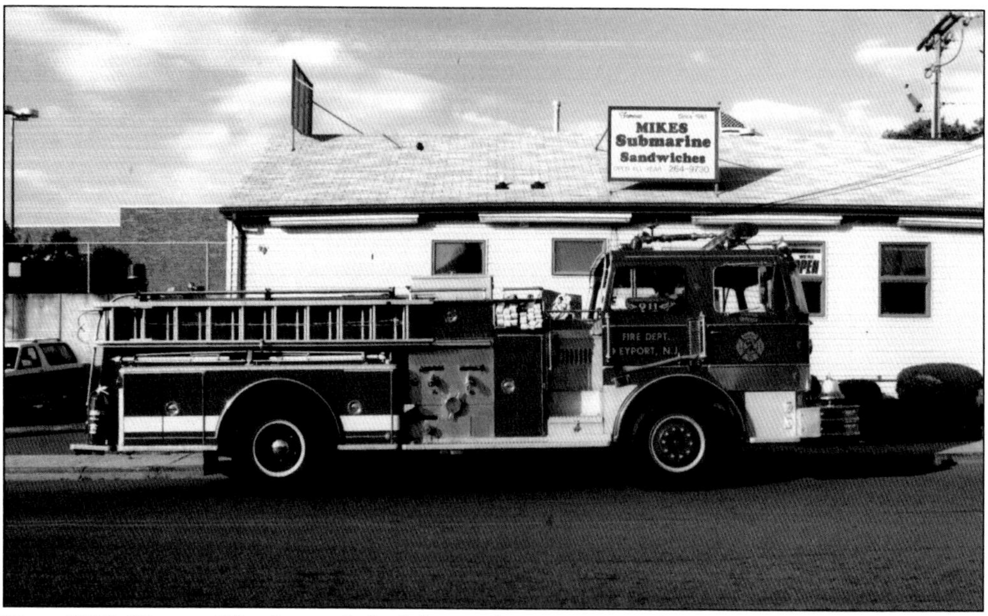

THE 1970 HAHN. The borough had been planning on replacing both the Raritan Hose Company and Keyport Fire Patrol trucks for some time when the 1972 Hahn was placed out of service. In an effort to continue the Keyport Fire Patrol's mission uninterrupted, Chief Anthony Vecchio procured this used Hahn pumper from the Oakhurst Fire Department. The truck was a standby for the patrol until a new apparatus could be purchased and constructed for them.

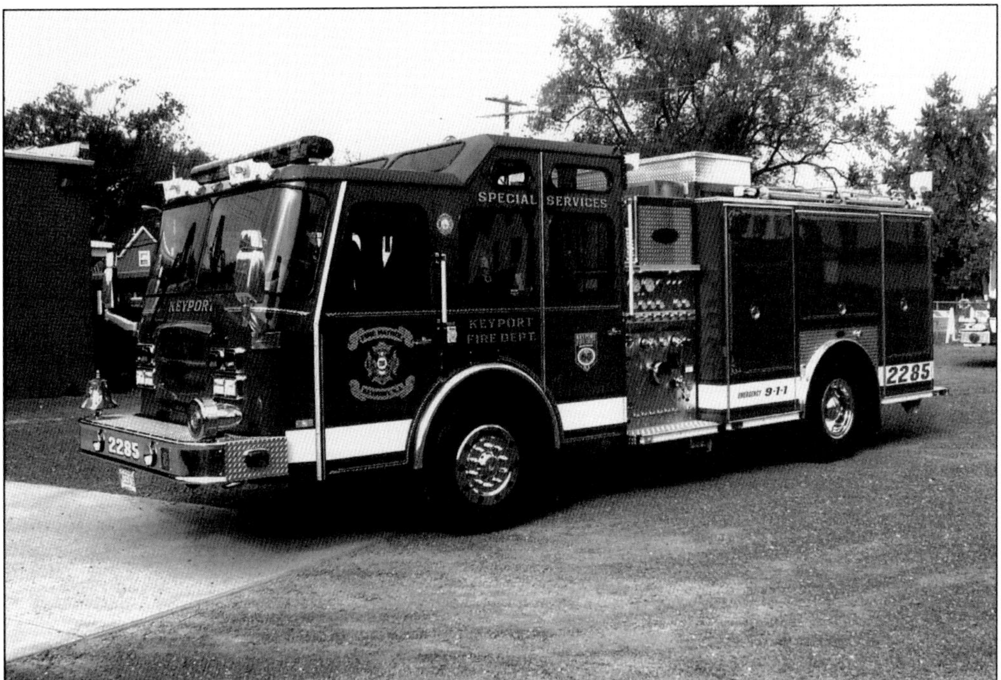

The 2003 E-One. Keyport Fire Patrol proudly received this Florida built Emergency One pumper in 2003. This unit became the sixth motorized unit of this company, the first having been delivered in 1925. This apparatus is equipped with a 1,250-gallon-per-minute pump, 500 gallons of water, and lighting, salvage, and traffic control equipment.

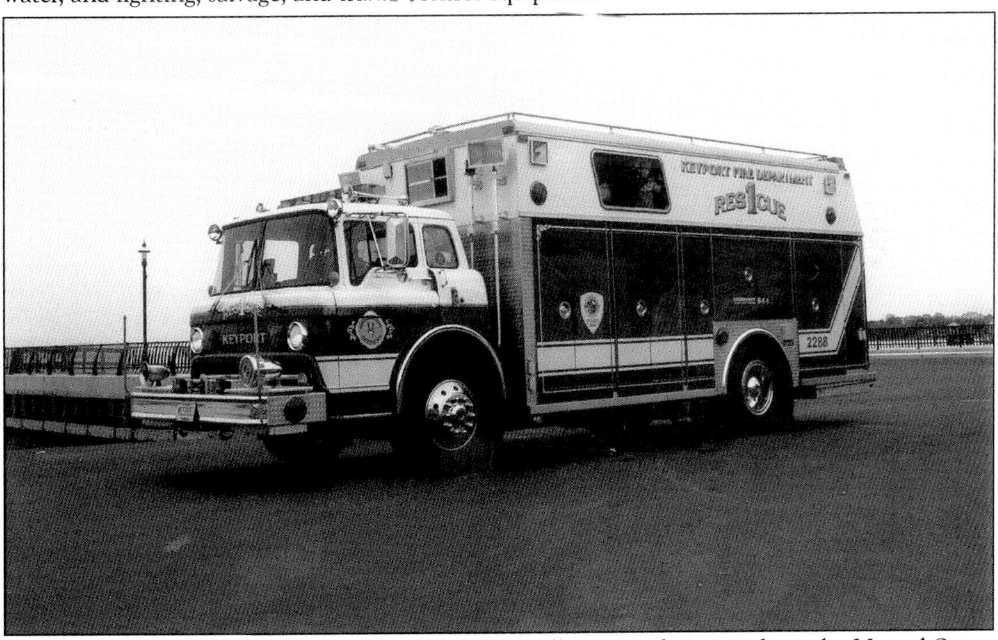

The 1988 Rescue Truck. In 2003, the KFD proudly received a grant from the United States government to purchase and equip a rescue unit. This 1988 Ford/Saulsbury was purchased from the Plainsboro Fire Company and was placed in service in 2003. It is staffed by members from the seven companies who are trained in its operation and specialized equipment.

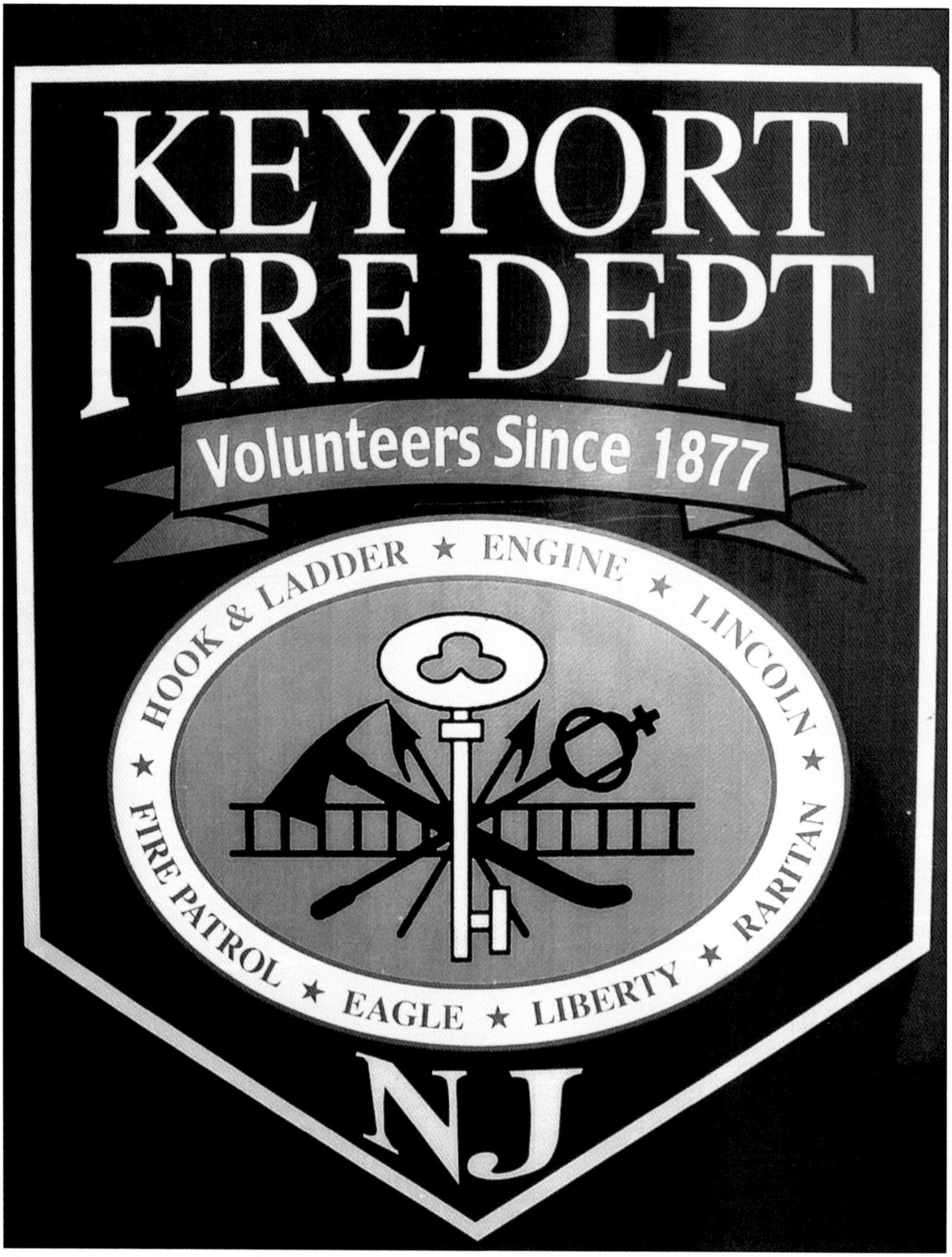

EMBLEM OF THE KFD. In preparation for the 125th anniversary of the department, it was decided to develop a new patch that would better signify the makeup of the seven companies. This patch was unveiled in 2001, and by the 125th anniversary, all uniforms were displaying the new emblem.

Four

TRAINING

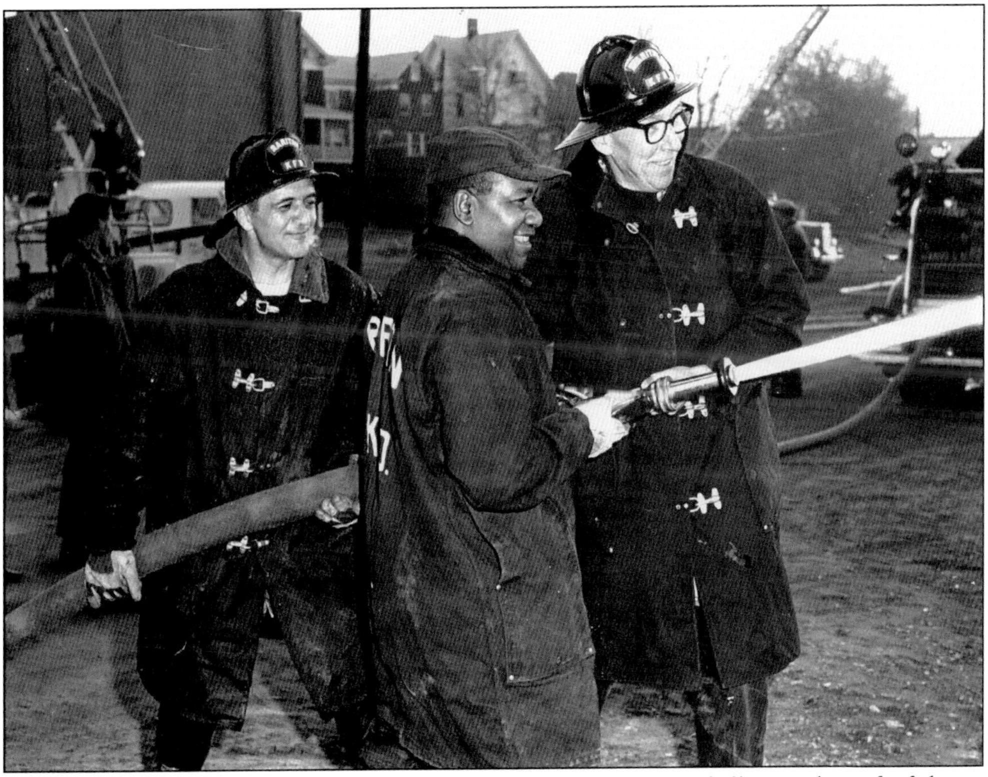

DEPARTMENT DRILL. Volunteer firefighters spend countless hours drilling to keep firefighting skills honed to perfection. This drill held on November 24, 1968, was attended by the home department as well as Union Beach, Cliffwood, and Matawan Borough and Township companies. Seen here from left to right, Leo McCann, George Washington, and Art Collins man the big two-and-a-half-inch hose line off the Raritan Hose Company pumper. The men are wearing low-profile metal fire helmets, which were in vogue until the early 1970s. (Photograph by James Atkins.)

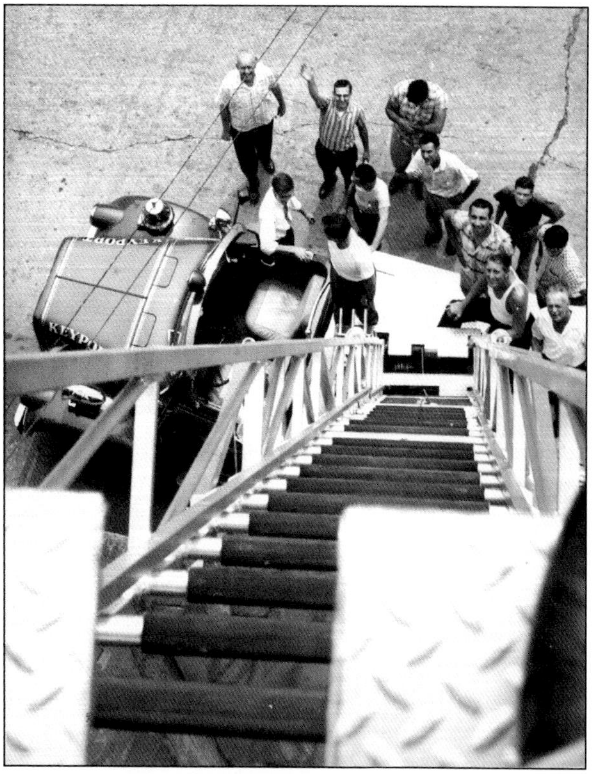

Looking Up and Looking Down. The capability of Keyport's firefighting force was greatly enhanced by the delivery of the aerial ladder truck in 1958. Up until this time, all ladders had to be hand-raised into position, a tedious and very dangerous job. With the arrival of the hydraulic ladder, rescues and hose lines could be set up and executed quickly.

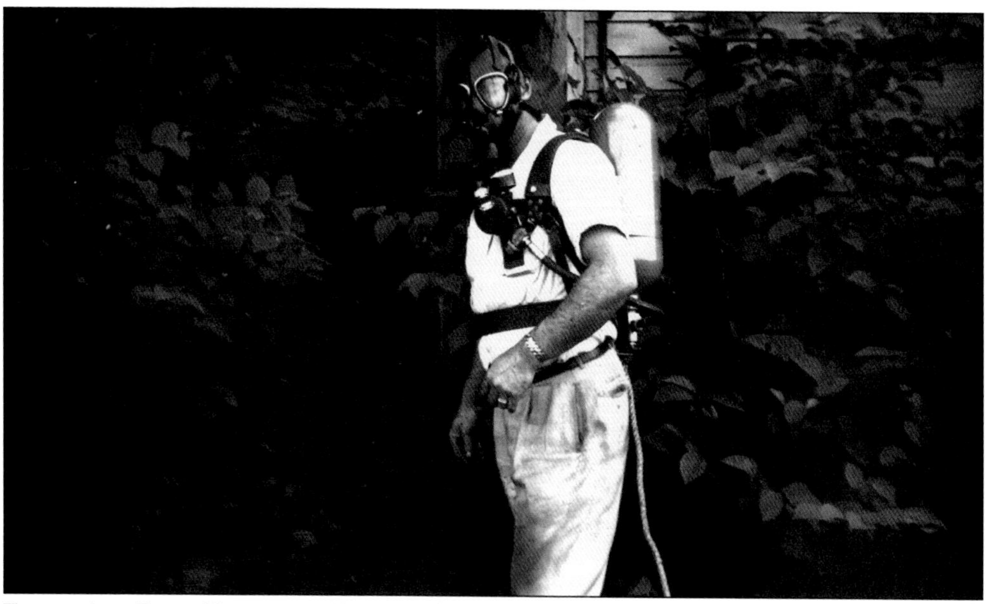

Early Air Pack Training. Keyport placed in service some of the earliest self-contained breathing apparatus (SCBA) in Monmouth County in late 1949. This assembly of pressurized air within a steel cylinder, regulator, and sealed mask all combined for a revolutionary advance in firefighting ability. For the first time, a firefighter could enter a smoke-charged atmosphere and quickly rescue trapped persons or extinguish the fire before it grew in intensity. Today, a half century later, nearly the same design is in use.

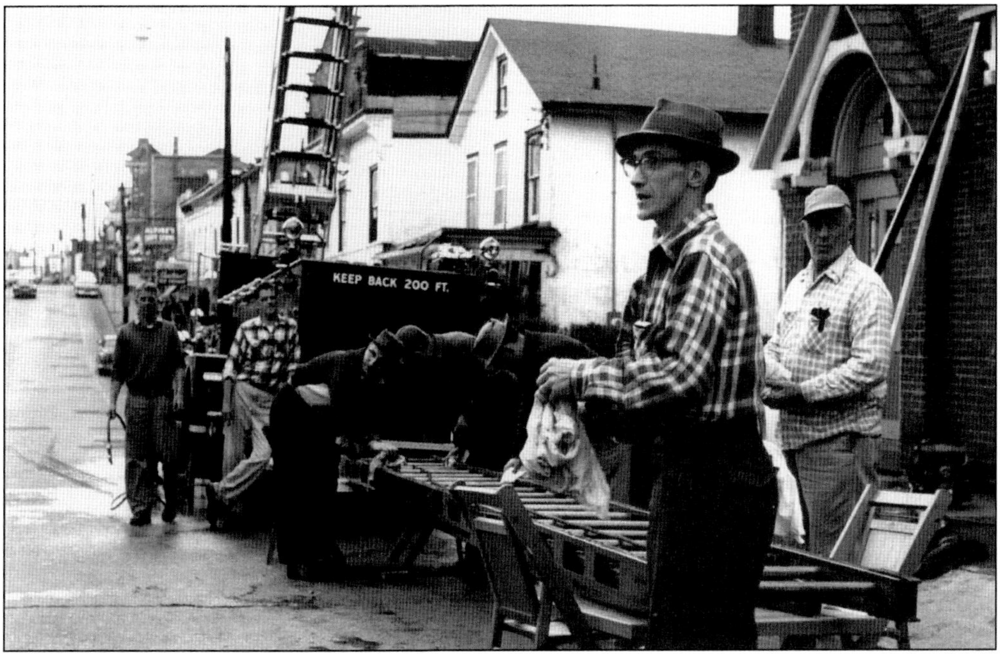

INSPECTION, C. 1961. Firefighters do not just fight fires. It is up to every volunteer to maintain each and every piece of fire equipment to ensure it is in proper working order. In this 1961 photograph, Bill Phillip and members of the Keyport Hook and Ladder Company remove dirt and refinish the ladders for an upcoming inspection. (Courtesy Keyport Hook and Ladder Company.)

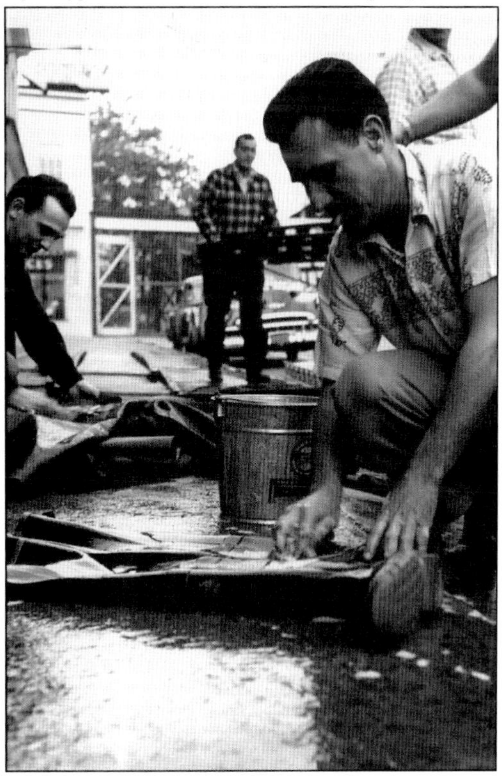

HOOK AND LADDER MEN HARD AT WORK. Each year, all borough-owned fire equipment is reviewed and inspected by the mayor, council, and visiting fire chiefs. This annual inspection has taken place each year since 1891, and firefighters work for over a week each year in preparation for it. No part of the apparatus will go unpolished, and each company will compete for "best appearing company" in the most friendly of rivalries. Members including Bob Rosato, Leroy Aumack, and future chief Harvey Ross work to get the 1958 Seagrave in shape for inspection.

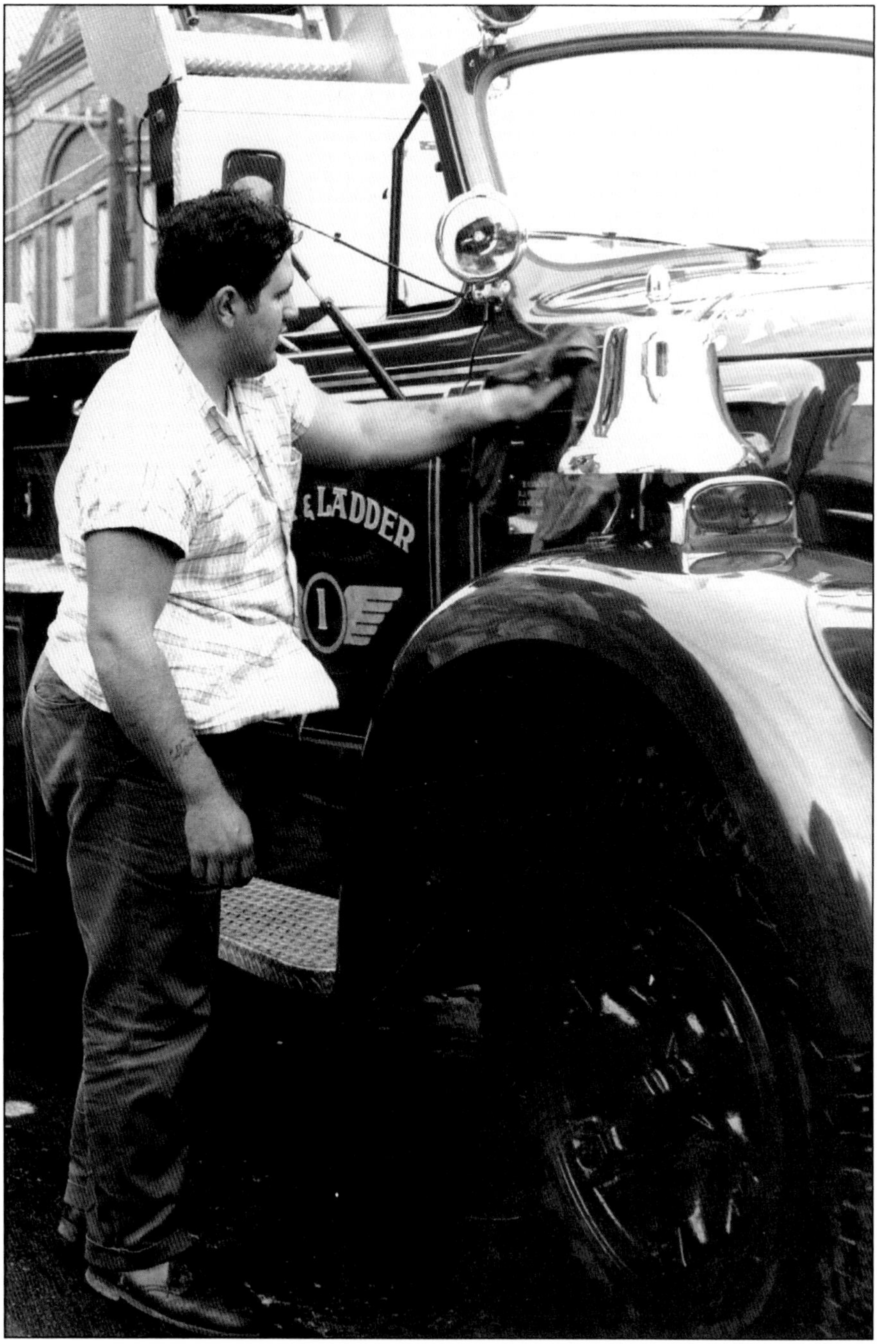

POLISHING THE SEAGRAVE. Firefighter Diodato is attending to every small crevice and detail in preparation for the big inspection day. The 1958 Seagrave last received such care for the 1987 inspection before the faithful old truck was sold at public auction.

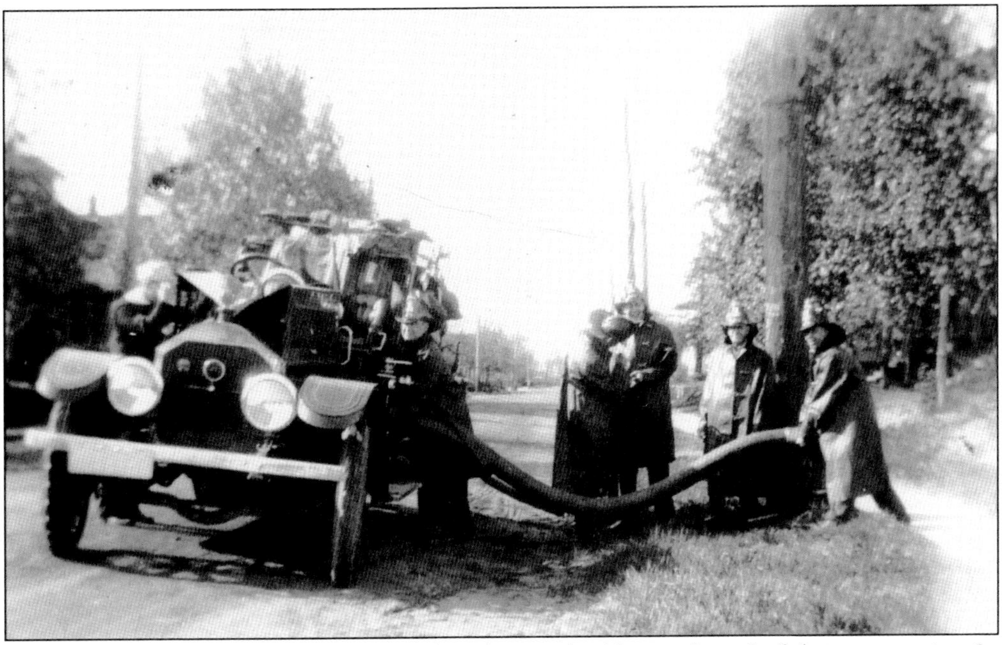

Pump Training 1921. The pumper truck is the true backbone of any firefighting operation. In 1921, Keyport Engine Company is seen working the 1919 LaFrance from a local fire hydrant. The engine's hood has been raised to allow the engine to stay cooler and prevent overheating while the engine pumps to its maximum 750-gallon-per-minute capacity.

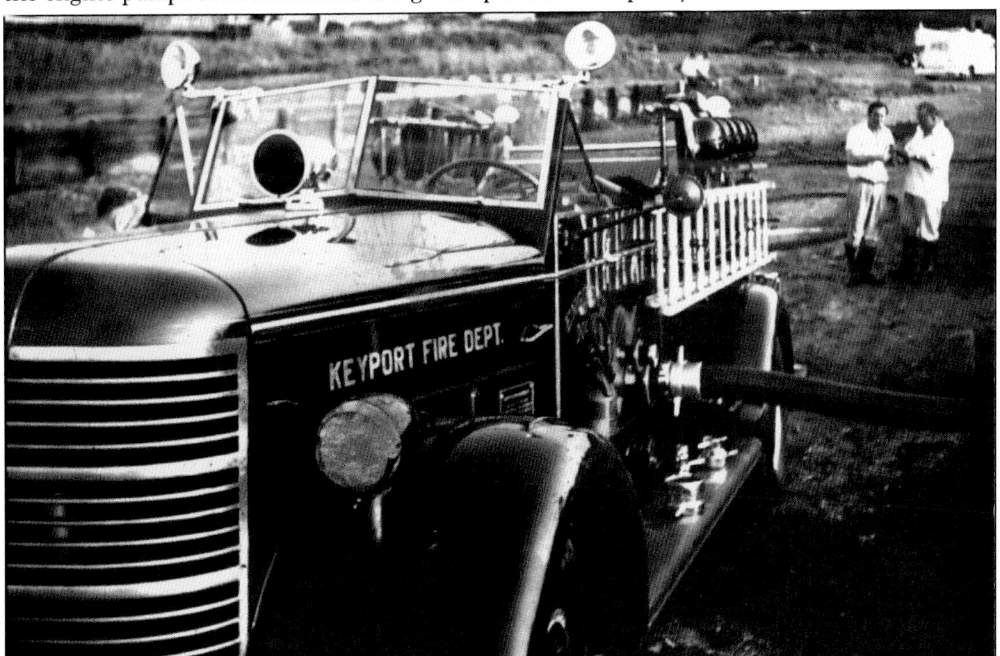

Pump Training, 1941. Just 20 years after the above photograph was taken, the members of Keyport Engine Company were captured drilling with their 1941 American LaFrance pumper. Fire engine drivers must operate a multitude of gauges, valves, discharges, and other equipment to guarantee that firefighters have an adequate water supply to get the job done.

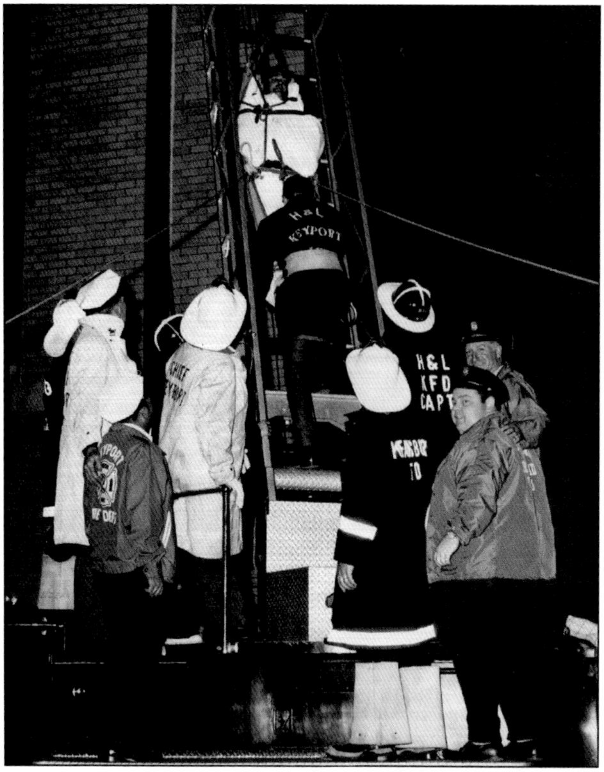

THE BIG RESCUE. One of the most difficult rescues is known as the "high angle rescue," which is when an injured person must be lowered from a building or high altitude by ropes or ladders. Keyport firefighters have been at the forefront of advanced training since the earliest days, and this drill night was no exception as an "injured" victim is lowered carefully from the school roof by hook and ladder men. (Courtesy Keyport Hook and Ladder Company.)

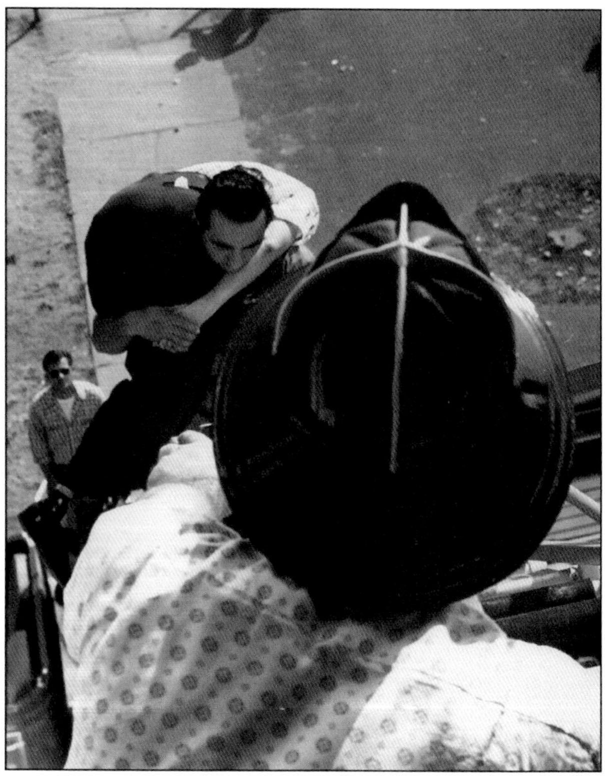

THE FIREMAN'S CARRY. Here a member of the Keyport Hook and Ladder Company carries an "unconscious victim" to the ground via the aerial ladder. The fireman's carry is a difficult maneuver to perform successfully, and constant training is the only way to become proficient in its use. Today it is highly discouraged to use live "victims," just in case.

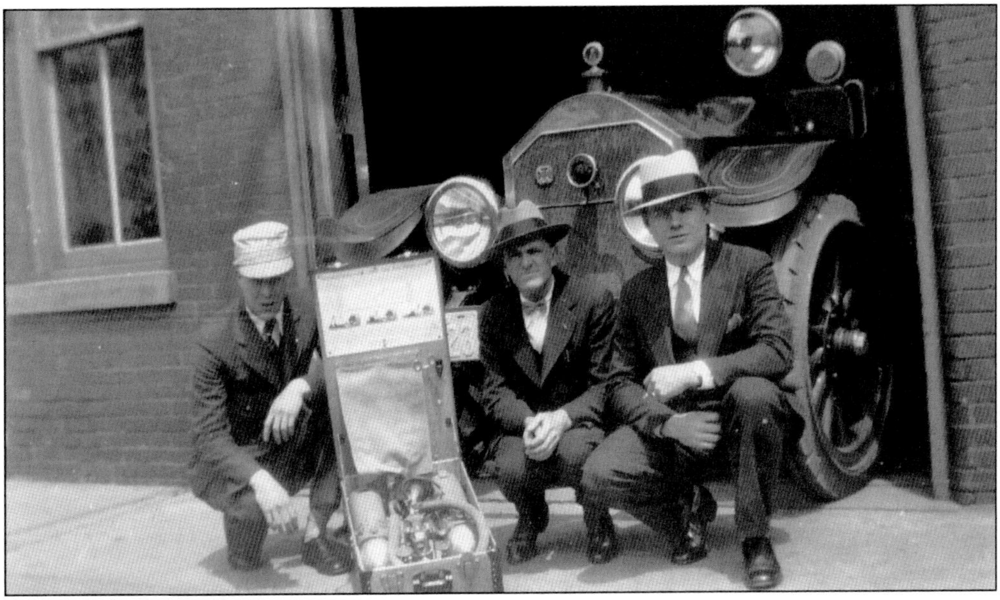

EARLY FIRST AID WORK. Prior to 1926, injured persons had to wait to be brought to local doctors' offices for treatment, sometimes with less-than-desirable results. It was during this time that the members of the Keyport Engine Company formed a first aid group and purchased a "lung motor," which was a form of an early respirator. During its first few years, countless people were revived from drowning and other emergencies by this machine, and the Keyport First Aid Squad was born in 1927. Soon each man graduated from the Red Cross first aid school, and calls for service grew quickly throughout the area. By 1929, the Keyport First Aid Squad had purchased an ambulance and today continues lifesaving work as an all-volunteer unit answering more than 1,200 calls annually.

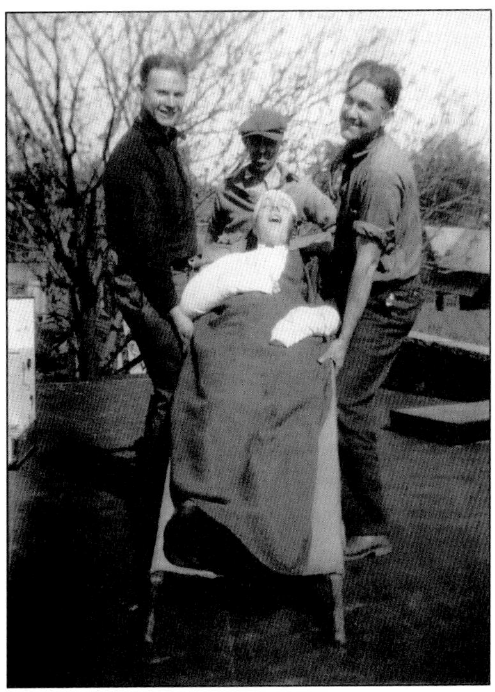

JUST CLOWNING AROUND, C. 1926. Members have a little fun while the "patient" acts out agony. This photograph was taken on the roof of the old Main Street firehouse of the Keyport Engine Company.

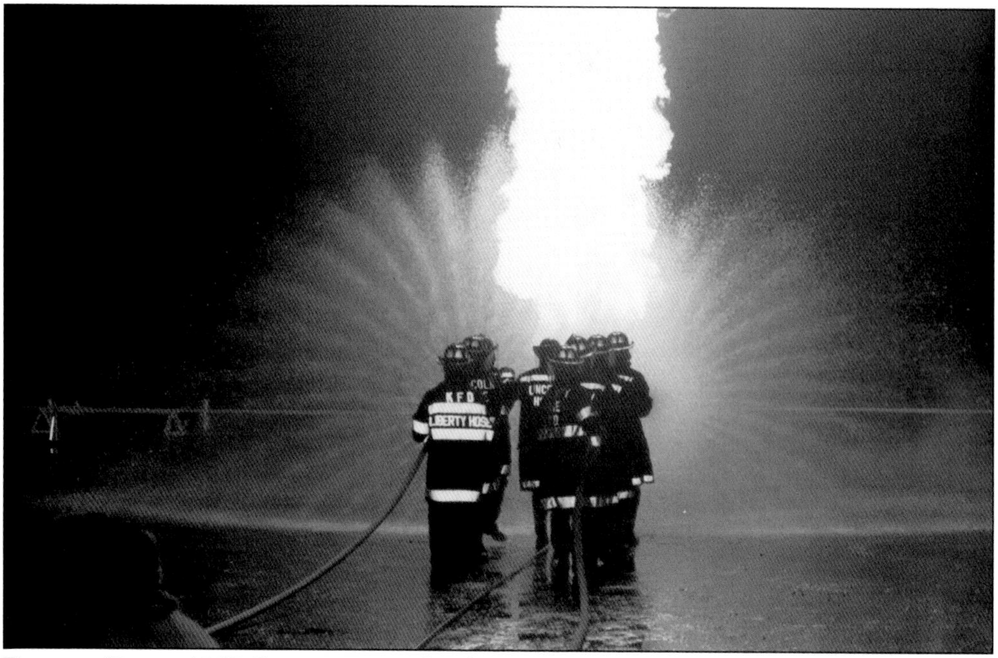

PROPANE DRILL, 1987. For many years, local gas companies would provide specialized training in liquefied petroleum gas fires for area firefighters. For many years, Keyport played host to this event, held along American Legion Drive at the old fairgrounds. Firefighters would advance the hose line, using the cooling properties as a shield, while the lead man would secure the gas valve and allow the fire to go out.

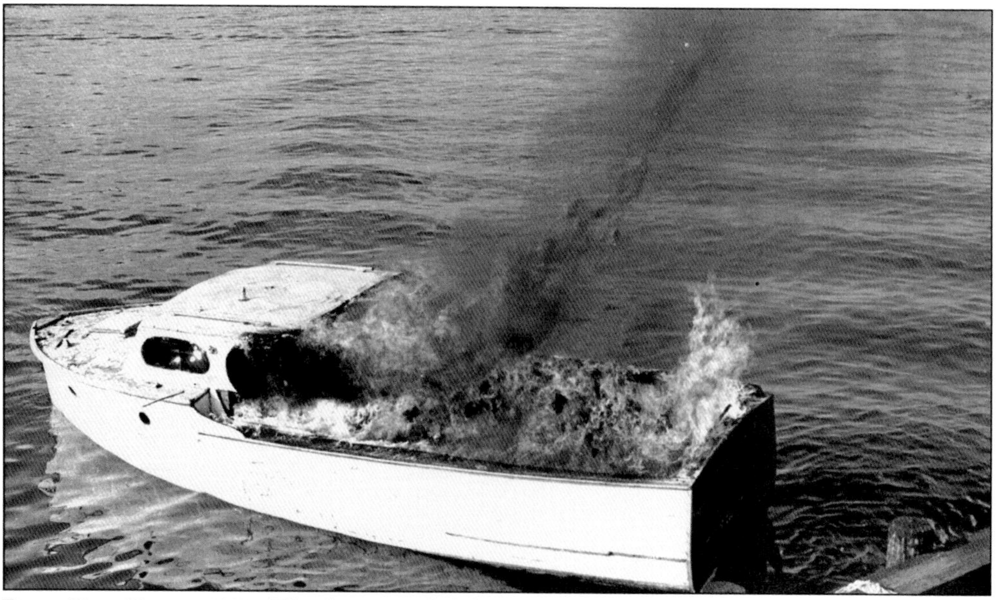

BOAT FIRE DRILL, 1970. The first recorded fire on a marine vessel occurred in 1893 when the steamer *Minnie Cornell* was completely destroyed by fire at the Keyport wharf. Since that time, fires have destroyed many commercial and pleasure watercraft, with the worst occurring in 1979 when a boat exploded in the Matawan Creek, claiming one life. Today the fire department is often called with its fireboat to suppress fires.

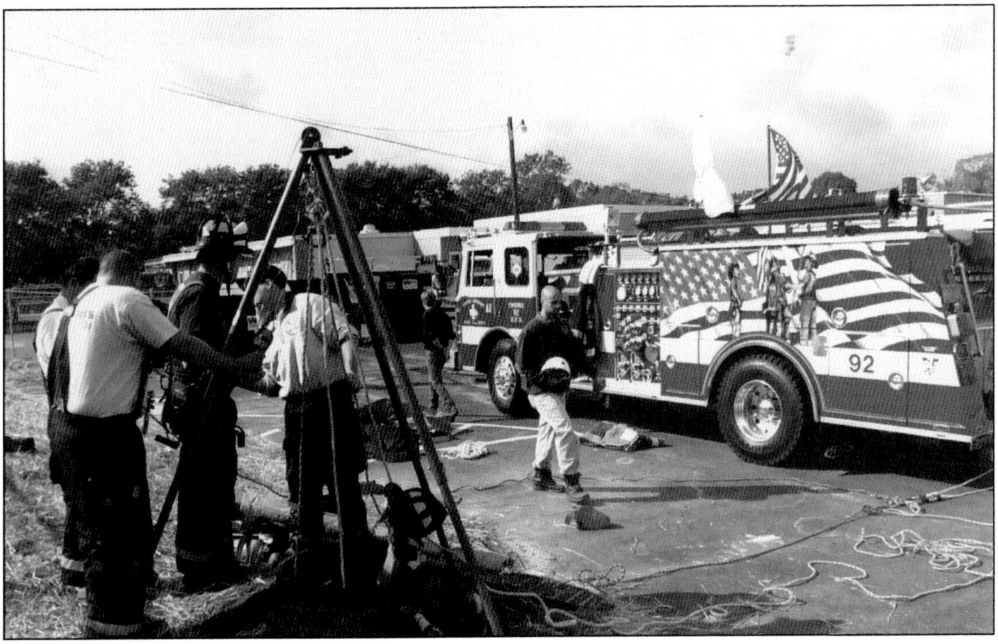

Confined Space Training, 2004. Keyport firefighters have invested time and money in learning the technical operations of confined space rescue. Having proper equipment to rescue a trapped worker in a manhole or deep passage requires skill and training to get the job done. This operation is very manpower intensive, and often up to 10 rescuers can be needed to save just one victim.

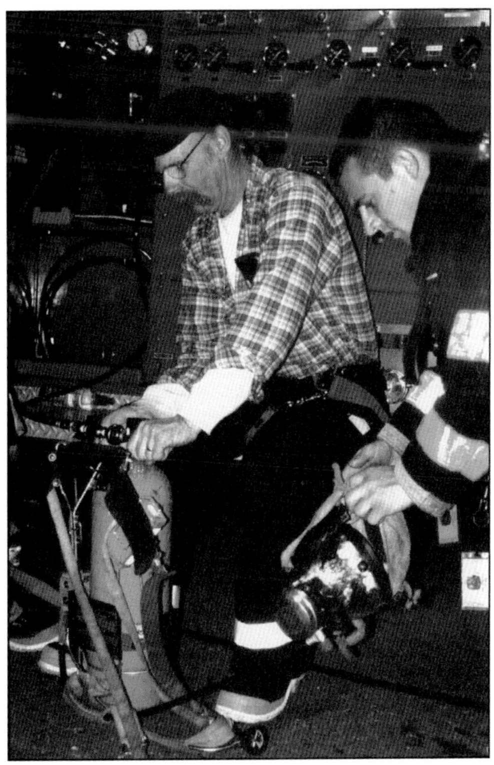

SCBA Refilling, 1998. Firefighters' SCBA cylinders allow them about 30 minutes of working time within a fire, before needing to be replenished. It has been the job of the Keyport Fire Patrol to refill air cylinders at fires since the early 1970s. Here ex-chief Ken Goble refills firefighter Walter Schaub's SCBA at a 1998 structural drill on Fulton Street.

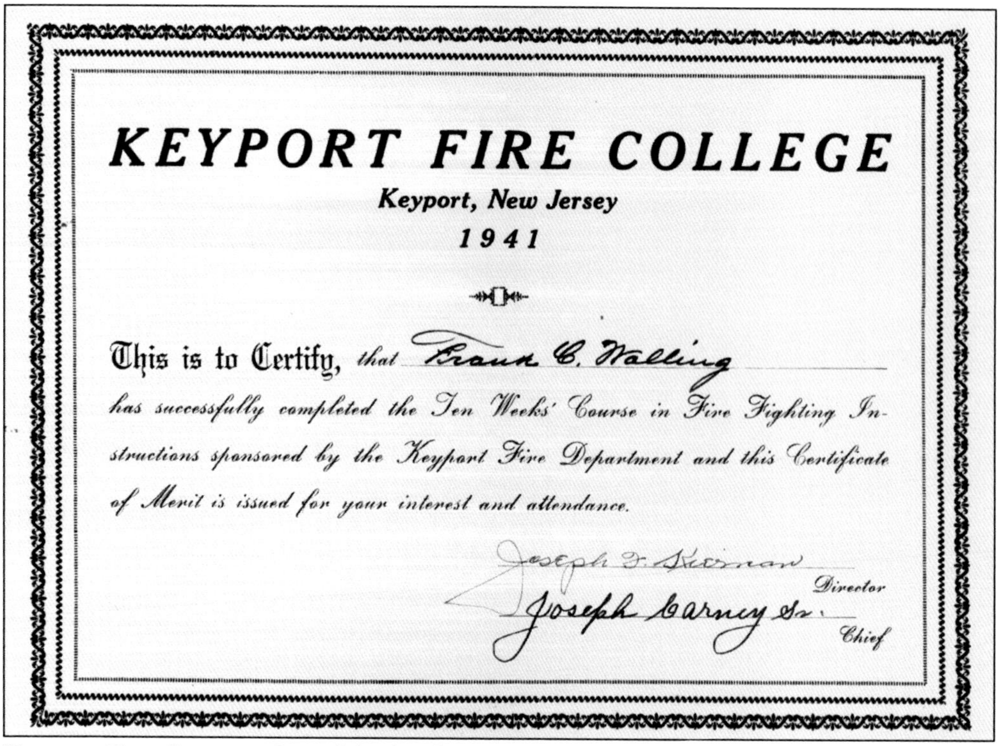

KEYPORT FIRE COLLEGE
Keyport, New Jersey
1941

This is to Certify, that Frank C. Walling
has successfully completed the Ten Weeks' Course in Fire Fighting Instructions sponsored by the Keyport Fire Department and this Certificate of Merit is issued for your interest and attendance.

Joseph F. Kiernan
Director
Joseph Carney Sr.
Chief

KEYPORT FIRE COLLEGE. One of the first fire service development courses was begun in Keyport in 1941. Local chiefs saw the need for advanced training to keep abreast of modern firefighting techniques, and soon the modern fire college was born. Instructors from Jersey City, Newark, and New York City were brought in to teach modern tactics to local firefighters, giving them an edge on a common foe—fire.

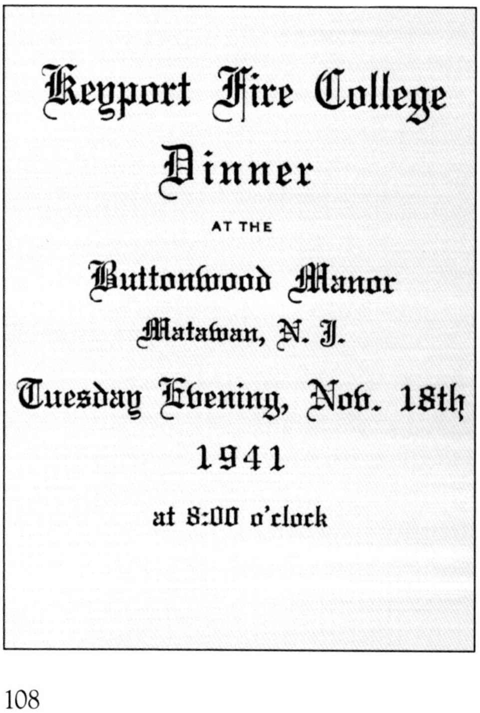

Keyport Fire College
Dinner
AT THE
Buttonwood Manor
Matawan, N. J.
Tuesday Evening, Nov. 18th
1941
at 8:00 o'clock

KEYPORT FIRE COLLEGE CERTIFICATE. Over the years, hundreds of firefighters from Monmouth County have attended the fire college classes and seminars. Such training formed the foundation of the Monmouth County Fire Academy, which was begun in the early 1970s in Howell Township.

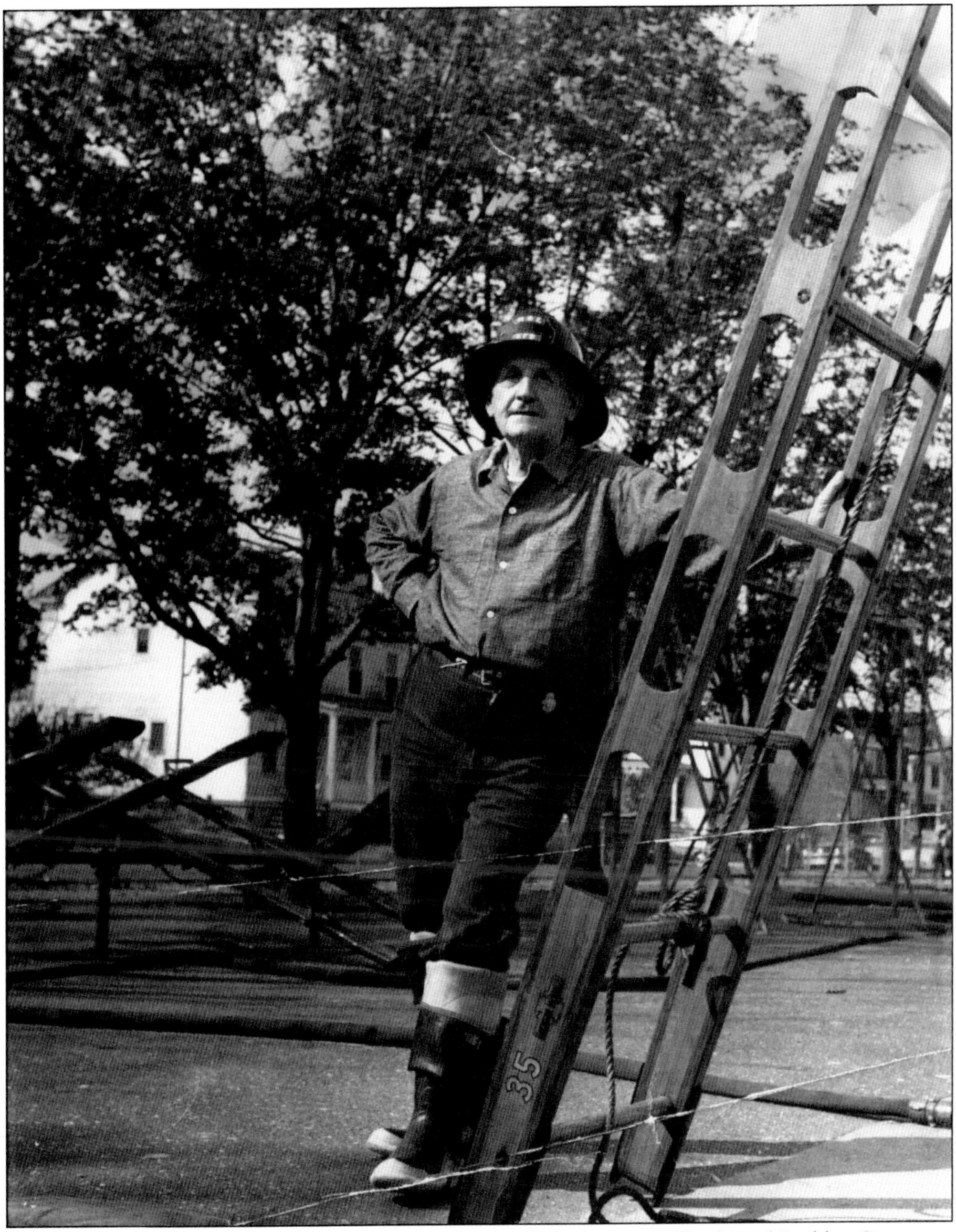

CLAUDE RAYNOR SR., 1950. Claude Raynor joined the Keyport Hook and Ladder Company in 1918 and spent his first two years pulling the old "man killer" to fires by hand. He would then drive and operate the 1920 Mack, the 1934 Diamond T, and finally the 1958 Seagrave. His membership spanned over six decades and is among the longest in the department's history. (Courtesy Keyport Hook and Ladder Company.)

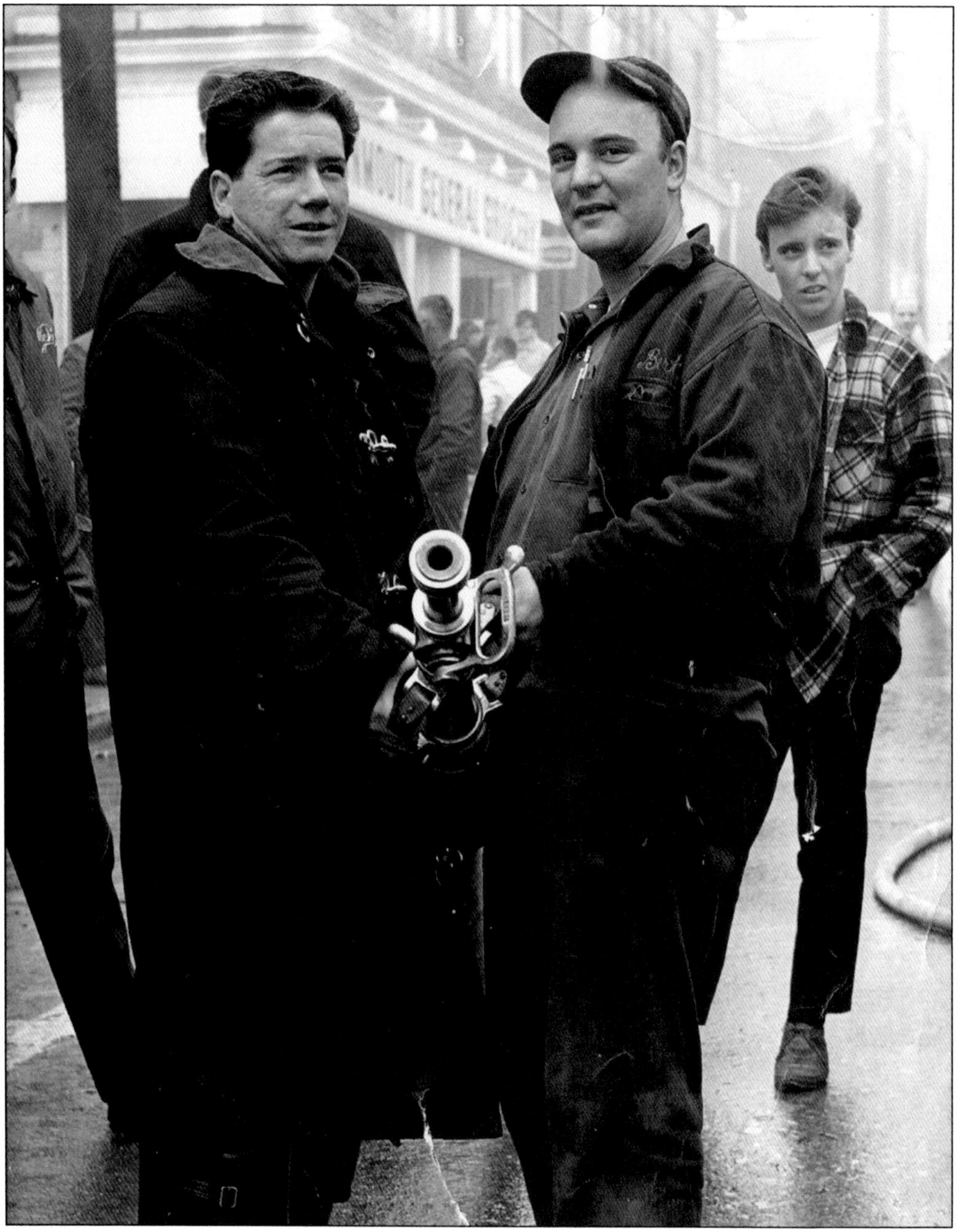

LIBERTY MEN, 1971. This 1971 drill in the center of the business district put Liberty Hose Company members Rich Brennan and Bert Morris to work on the old two-and-a-half-inch nozzle. Future Liberty Hose Company member and son of Bert Morris, Mark Morris looks on. These three men are part of a service that has had over 1,500 members since 1877, all of whom have served with no compensation for their often dangerous work. (Courtesy Raritan Hose Company.)

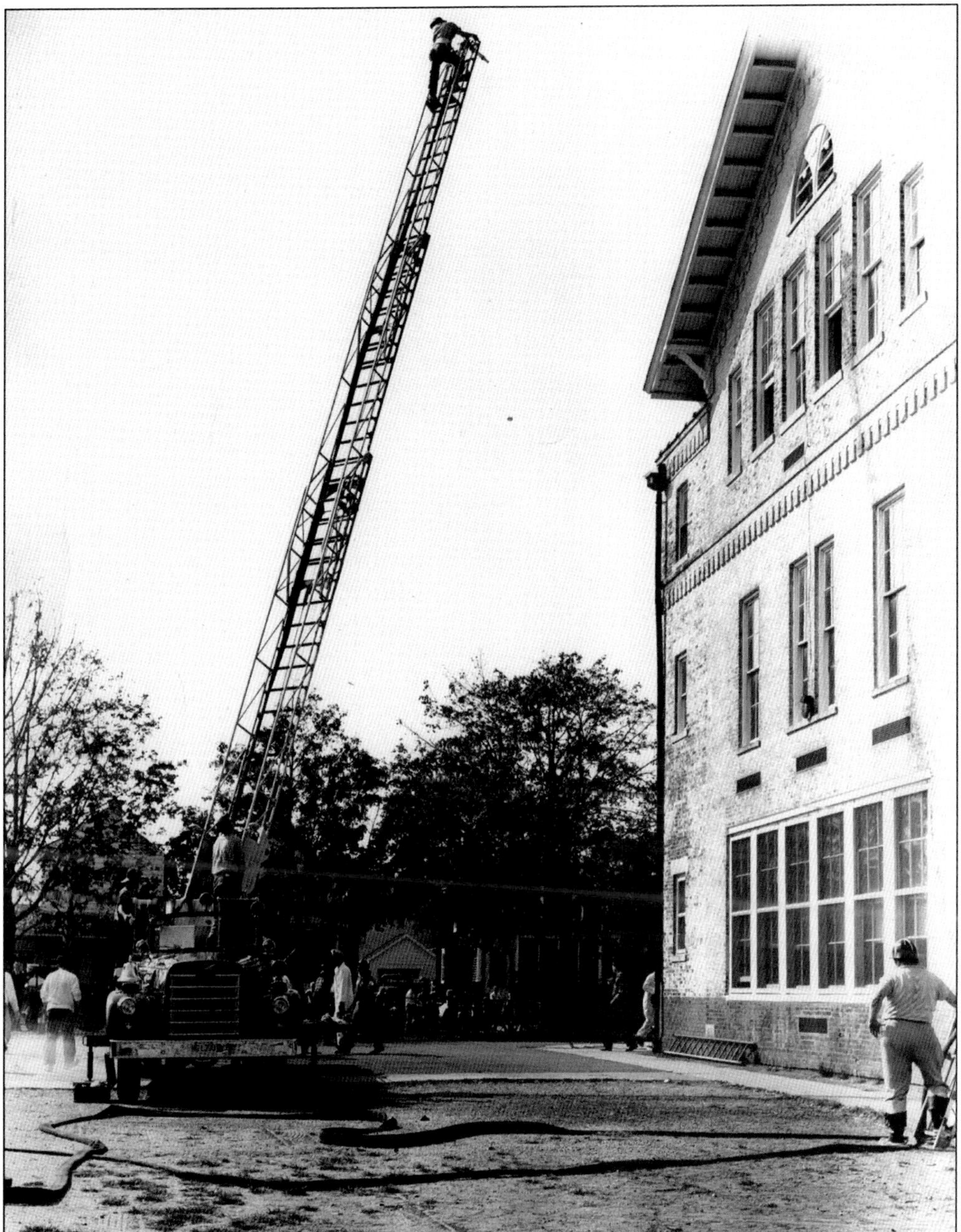

TRAINING AT THE OLD GRADED SCHOOL, 1966. Fire company members must dedicate themselves to the importance of training during their volunteer career. It is common for members to dedicate 50 hours or more a month to training, drilling, and attending fires and meetings, all done without pay. (Courtesy Keyport Hook and Ladder Company.)

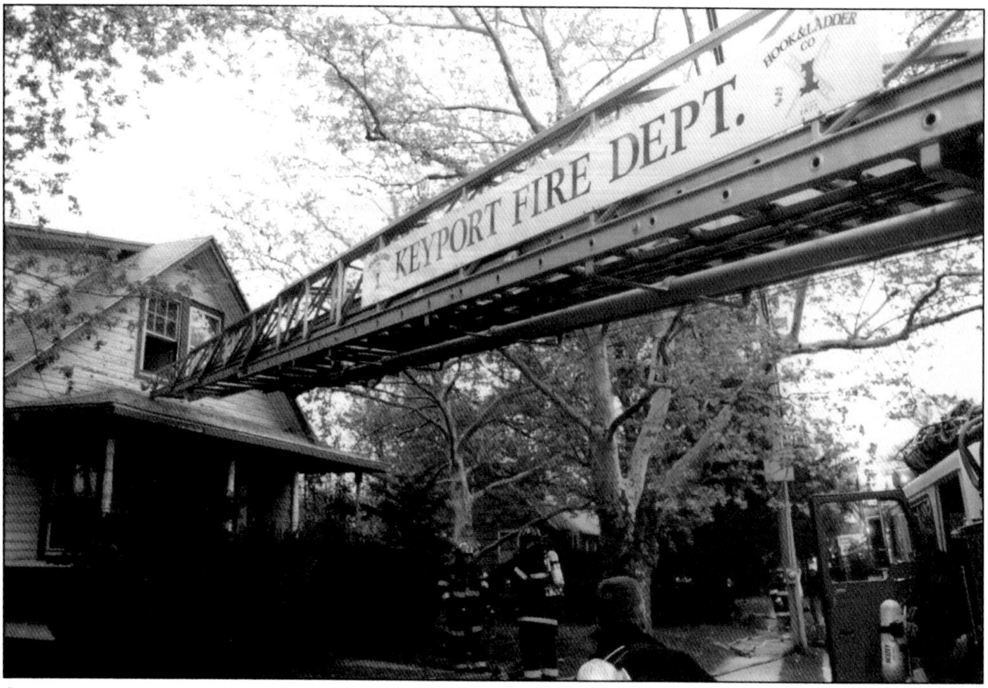

STRUCTURAL FIRE TRAINING, 2002. Often the fire department has the chance to acquire buildings that are scheduled to be demolished for training purposes. Firefighters, both seasoned and probationary, are able to practice skills in actual buildings, including roof ventilation, search and rescue operations, and hose line advancement.

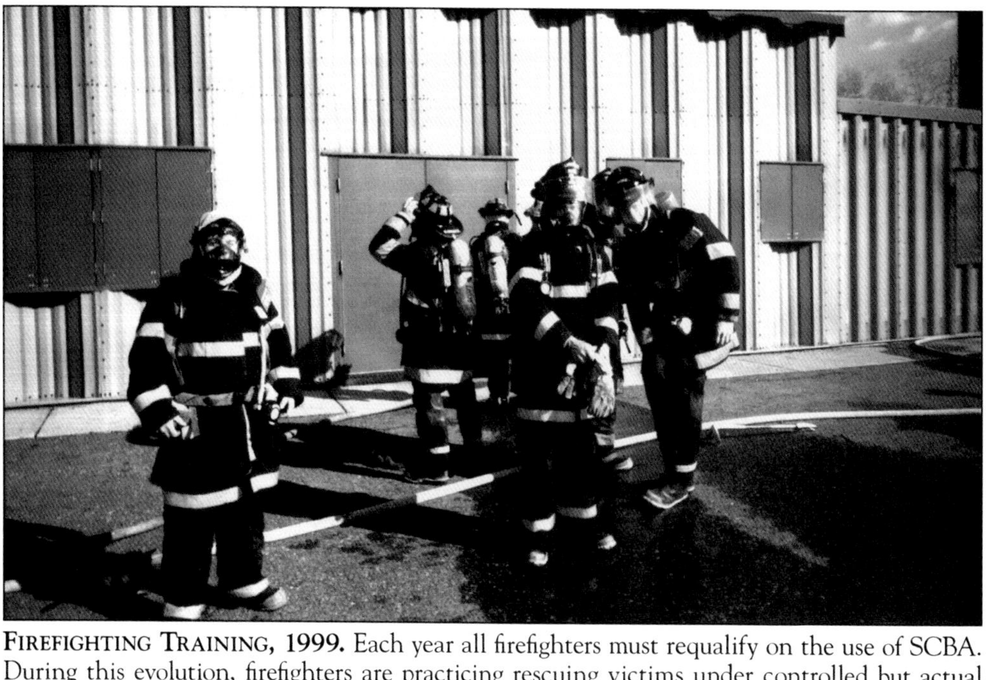

FIREFIGHTING TRAINING, 1999. Each year all firefighters must requalify on the use of SCBA. During this evolution, firefighters are practicing rescuing victims under controlled but actual fires in the Aberdeen Township smokehouse. Training such as this ensures that all volunteers are ready to save lives and property at all times.

Five

FIRST AID SQUAD

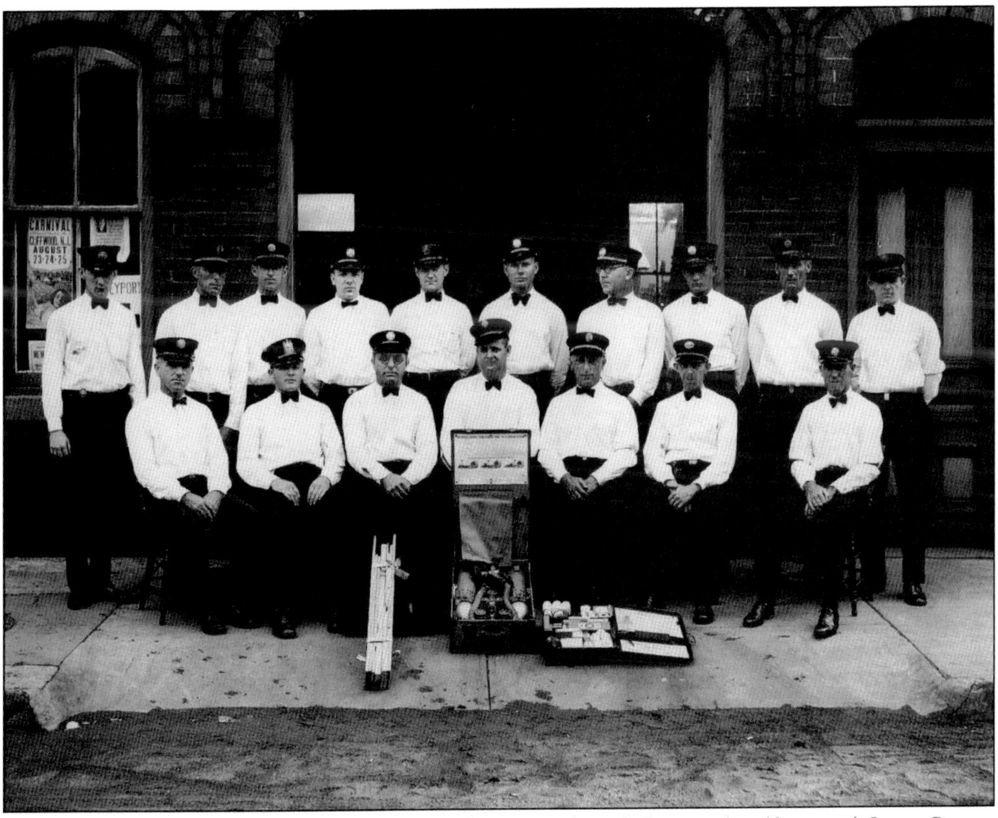

FIRST AID SQUAD MARCH 8, 1929. Seen here are, from left to right, (first row) Jesse Camp, Frank Van Pelt, Elwood Turlington, Edward Hopkins, Jesse Hywer, John Kelly, and John Haley; (second row) Robert Todt Jr., David Schanck, William Greene, Joseph Judson, Charles Cuttrell, John Dane, James Cadoo, Rufus Walling Jr., Edward Walters, and Edward Kelly. (Courtesy Keyport First Aid Squad.)

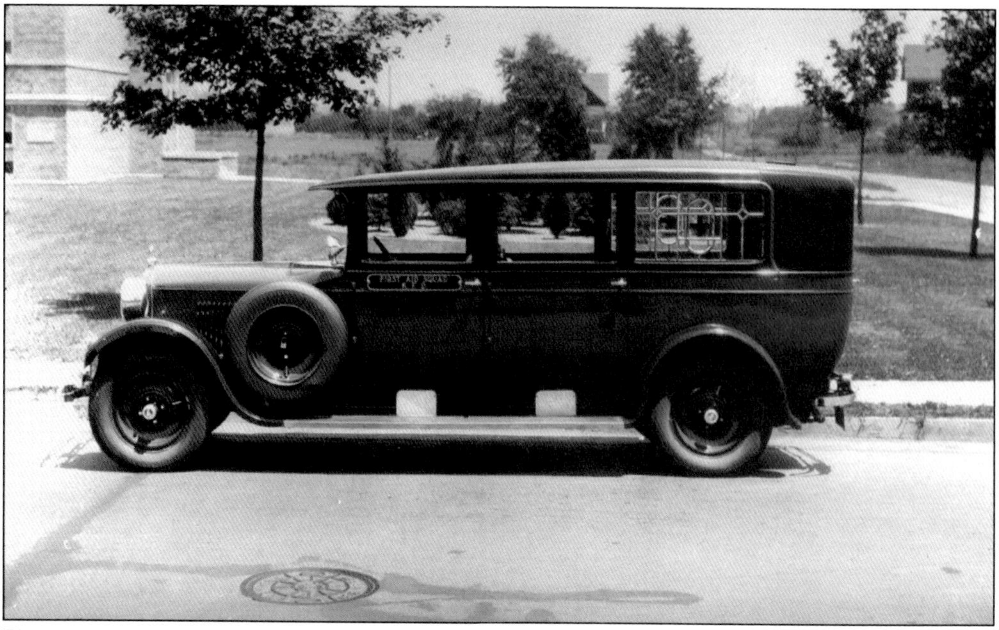

The 1929 Ambulance. The Keyport First Aid Squad placed this ambulance in service in 1928. Although it appeared to be little more than a dressed-up hearse, it was staffed with the equipment and tools needed to save lives. More importantly, the men who staffed the ambulance were trained by both the local Red Cross chapter and physicians to provide the first real pre-hospital emergency care in the state of New Jersey.

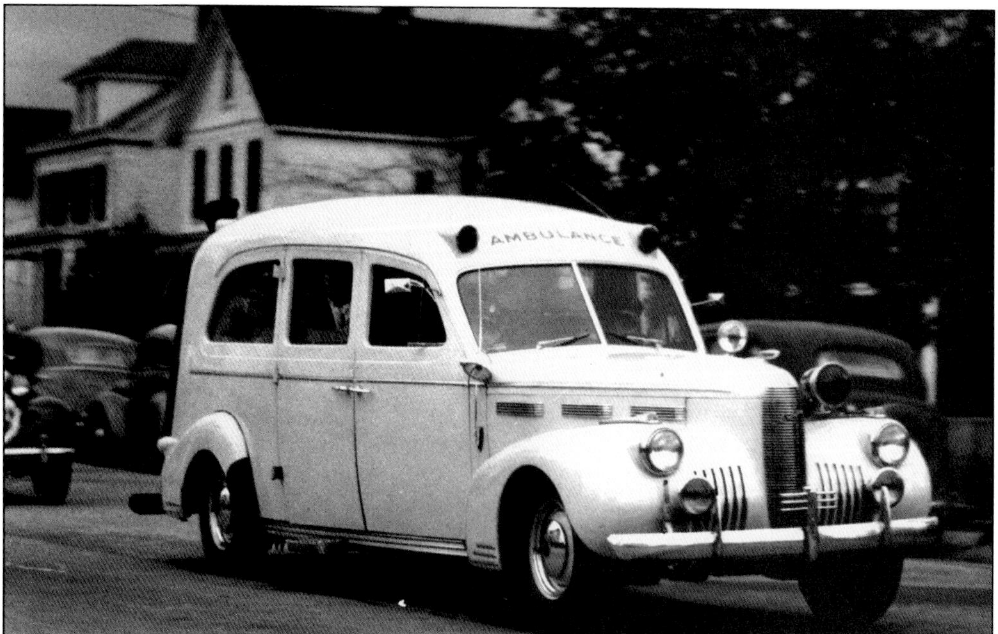

The 1940 LaSalle. Here is the second ambulance operated by the Keyport First Aid Squad on May 4, 1951. The Keyport First Aid Squad broke away from the fire department by the mid-1930s and became a separate entity. In 1974, the last Cadillac ambulance was purchased. Since that time all ambulances are built on customized heavy-duty van chassis.

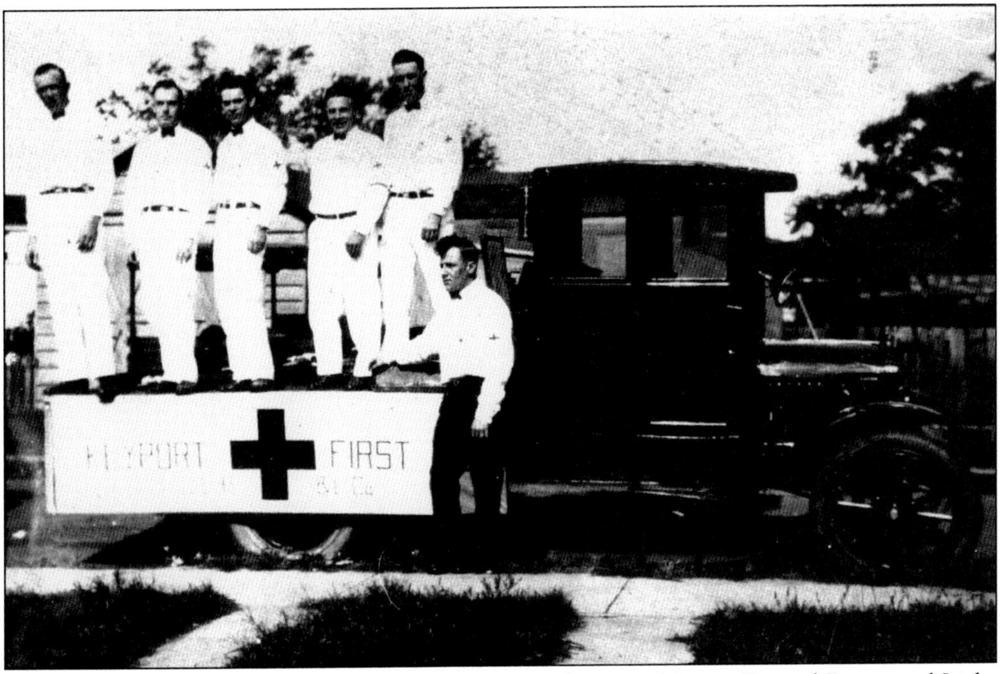

A 1931 Parade Photograph. Members pose on a borrowed Jersey Central Power and Light flatbed truck, assumed as such because the first aid banner has that electric company's name on it. The float is not overly adorned, but the pride is visible in their smiles.

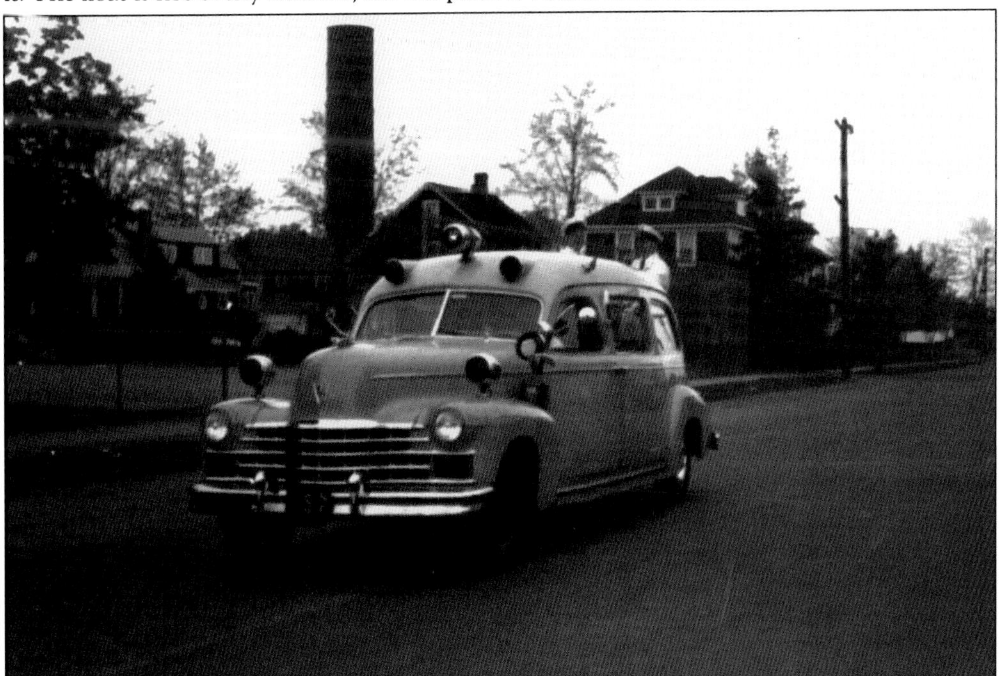

The 1946 Cadillac Ambulance. The Keyport First Aid Squad used a 1946 Cadillac to provide its services. The vehicle was well appointed with multiple warning lights. This photograph was taken on Maple Place looking east with the town's original 1893 water tower in the background.

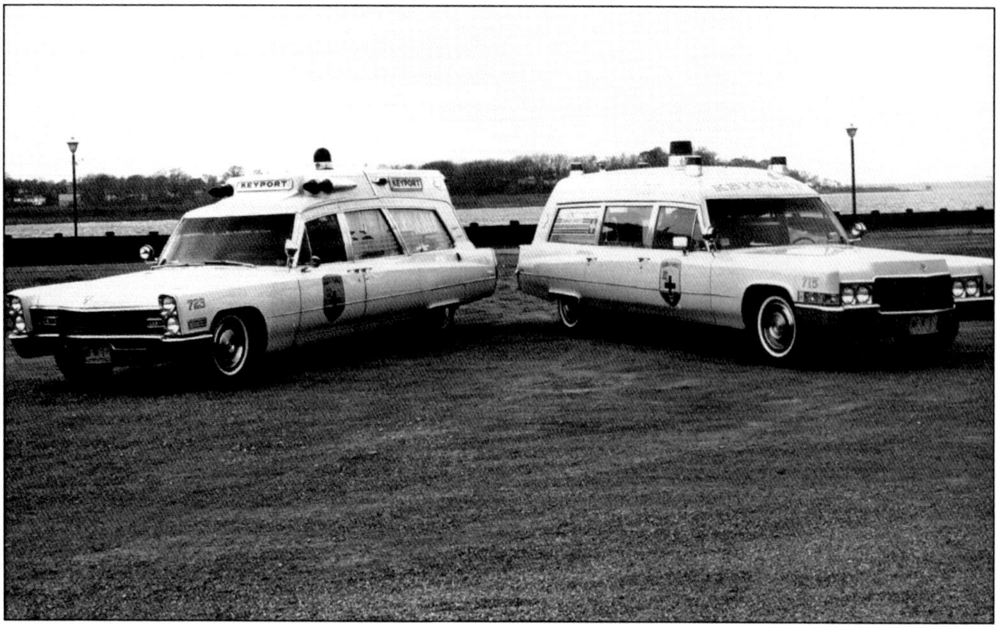

AND MORE CADILLACS. Located at Keyport's waterfront are a 1967 Cadillac Superior No. 723 (left) and No. 715, a 1969 Cadillac Meteor. Notice the different shapes of the emergency lights and the headlight orientation. (Courtesy Keyport First Aid Squad.)

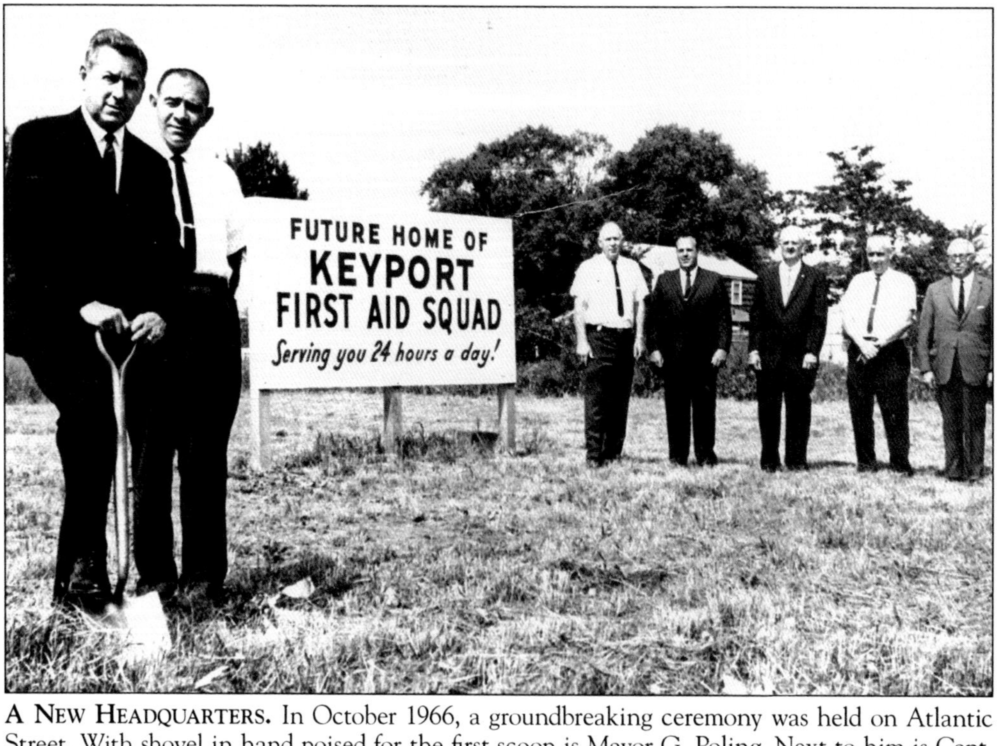

A NEW HEADQUARTERS. In October 1966, a groundbreaking ceremony was held on Atlantic Street. With shovel in hand poised for the first scoop is Mayor G. Poling. Next to him is Capt. Donald Miller, chairman C. Antrim, builder Ralph Seber, the oldest squad life member E. Hooking, president S. Heyer, and attorney H. Rothenberg. (Courtesy Keyport First Aid Squad.)

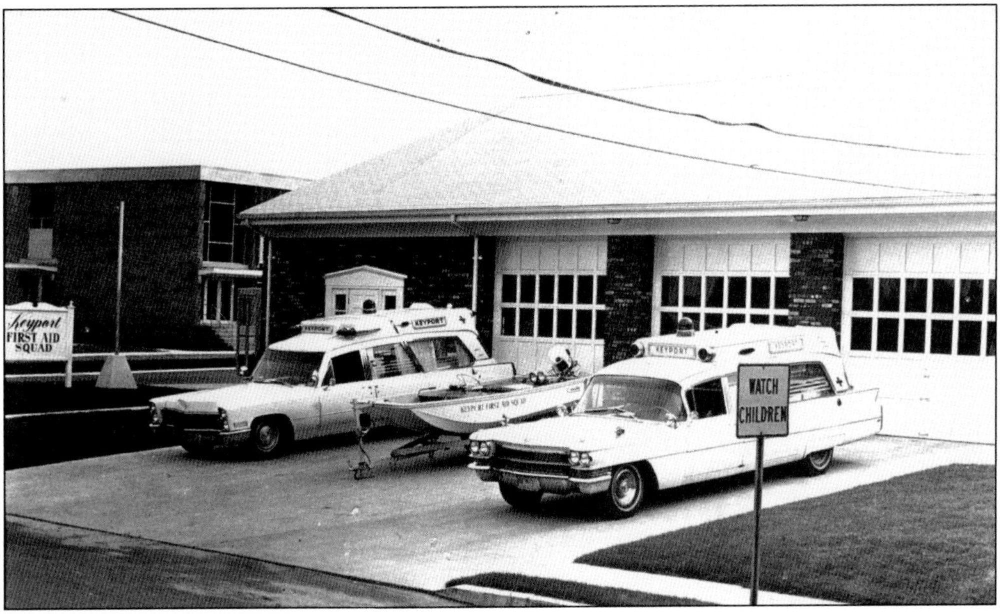

DEDICATION DAY. The new triple-bay building is completed. It was dedicated by invitation on October 1, 1967 at 1:00 p.m. Its address is, naturally, 1927 Atlantic Street. In front of the building is the 1967 Cadillac Superior (left). Next is the Keyport First Aid Squad rescue boat, a handy piece of equipment for a bayside town. The 1963 Cadillac Superior rounds out this happy occasion. The Keyport First Aid Squad has since expanded it quarters where it remains to this day.

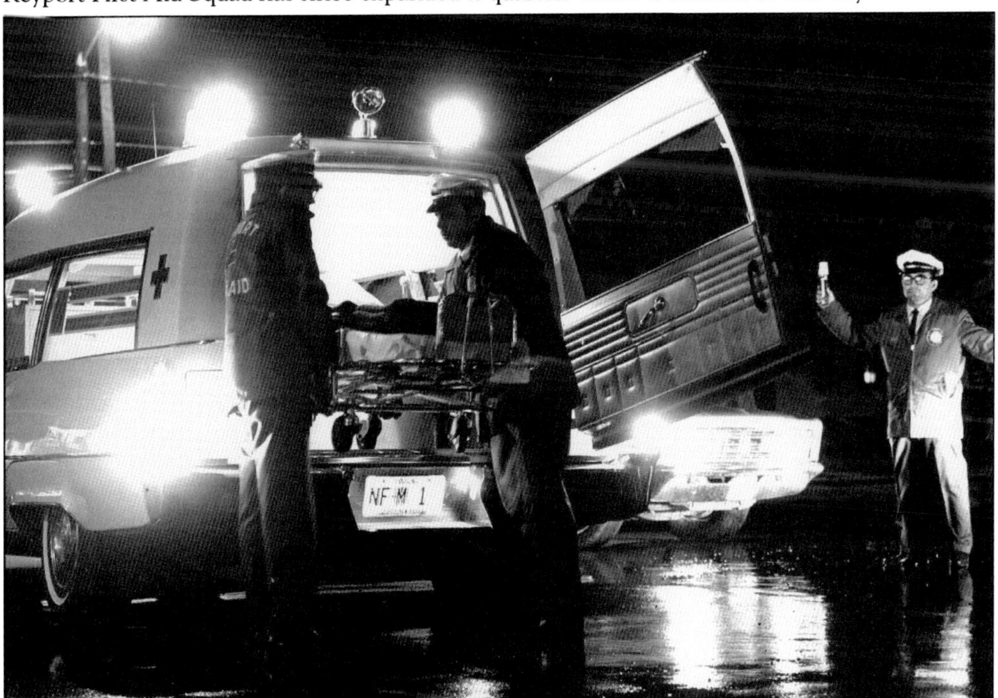

NIGHT MANEUVERS. Two members, one unidentified and Harvey Brandt (right), take the stretcher out of the rig while member Matt Ferris holds up traffic to keep a clear path.

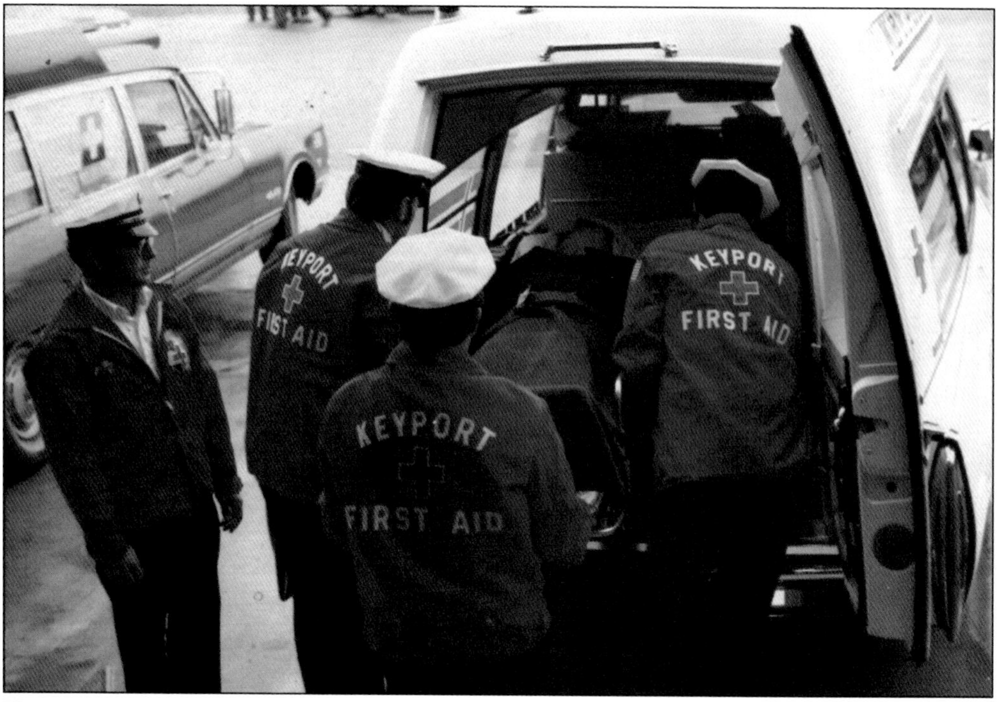

HELPING HANDS. Keyport First Aid Squad members train at a mutual aid drill at International Flavors and Fragrances of Union Beach and Hazlet in January 1975. From left to right are members Matt Ferris, Harris Smith, Harvey Brandt, and unidentified.

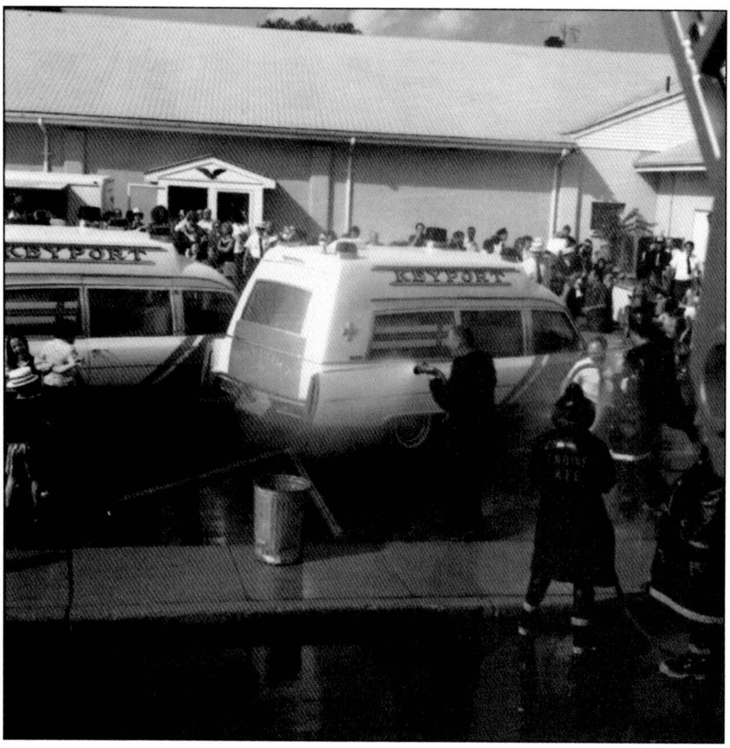

A NEW RIG. As with fire trucks, wetting down a new ambulance brings excitement. First it means one of the old rigs is retiring. The new rig will have more modern equipment and maybe a little more working room inside. These Cadillacs were extremely fast and nice to look at, but inside they were a bit small for working on victims.

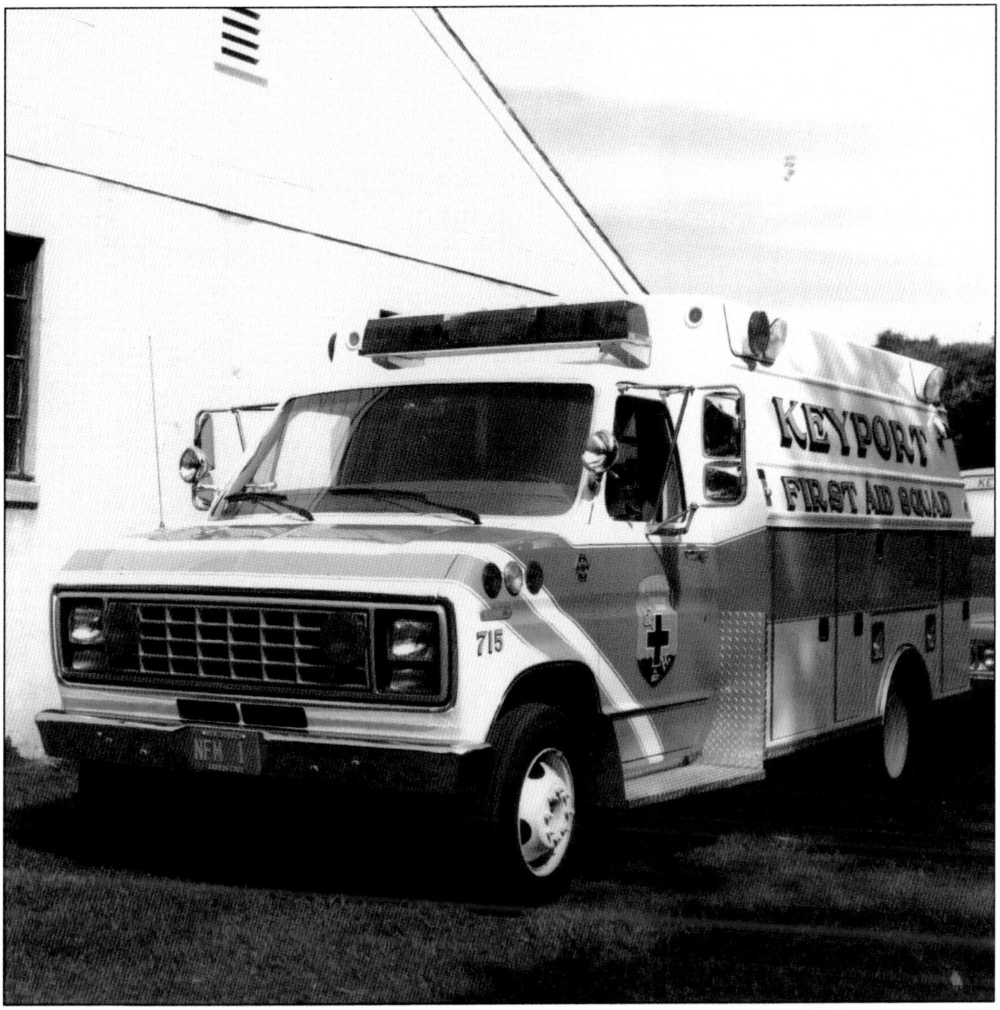

A CHANGE OF DESIGN AT LAST. In 1982, Keyport First Aid Squad purchased a Ford Braun, the first non-Cadillac. There is clearly more head and working room, side compartments, and heavier axle loadings. The unit's numbers are passed on to successive vehicles, this being No. 715. In September 1995, two new ambulances, dubbed "the Twins" were given the wet-down treatment as guests enjoyed hot dogs and hamburgers. These units received numbers 2256 and 2257 and currently continue to serve.

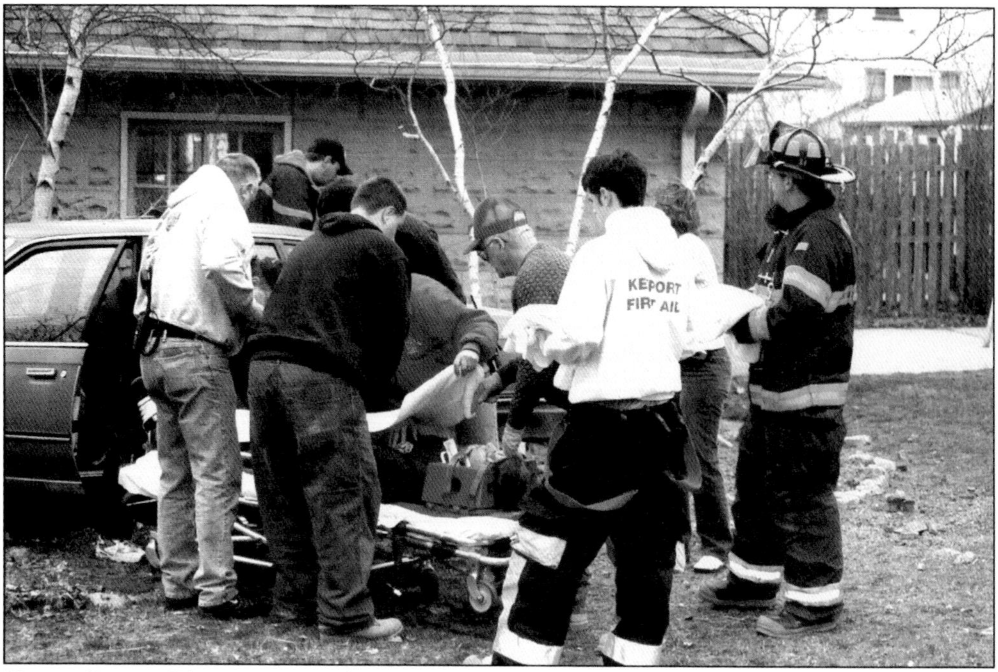

NOT A DRILL. This "car into a house call" resulted in minor injuries to the driver and some slight damage to the house. Seen here, from left to right, Rick Ely, Steve Adamski, Bob Poling, Nick Vecchio, and Joe Carson, in fire gear, tend to a victim with head trauma. Behind the car wearing a baseball hat is Rich Leswing VII.

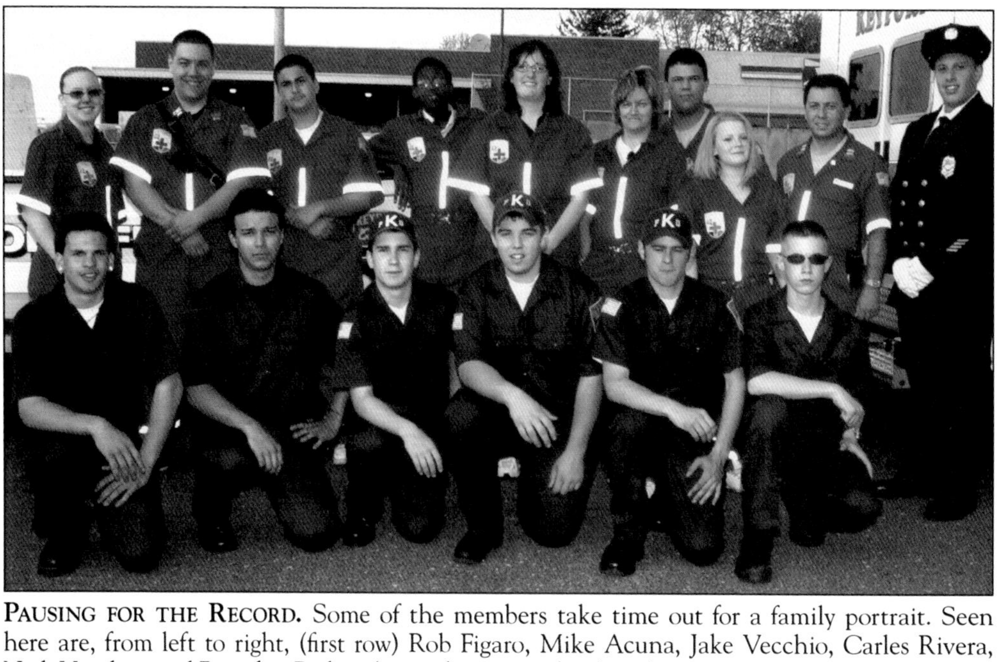

PAUSING FOR THE RECORD. Some of the members take time out for a family portrait. Seen here are, from left to right, (first row) Rob Figaro, Mike Acuna, Jake Vecchio, Carles Rivera, Nick Vecchio, and Brandon Parker; (second row, standing) Kathy Stewart, Steve Adamski, Sam, Wayne Ramadon, Ruth Grabowski (president), Lonajoyce Gallo, Phil Specchio, Natalie Ferm, Capt. Ken Krohe, and Phil Santiago.

Six

FIRE MUSEUM AND EDUCATION CENTER

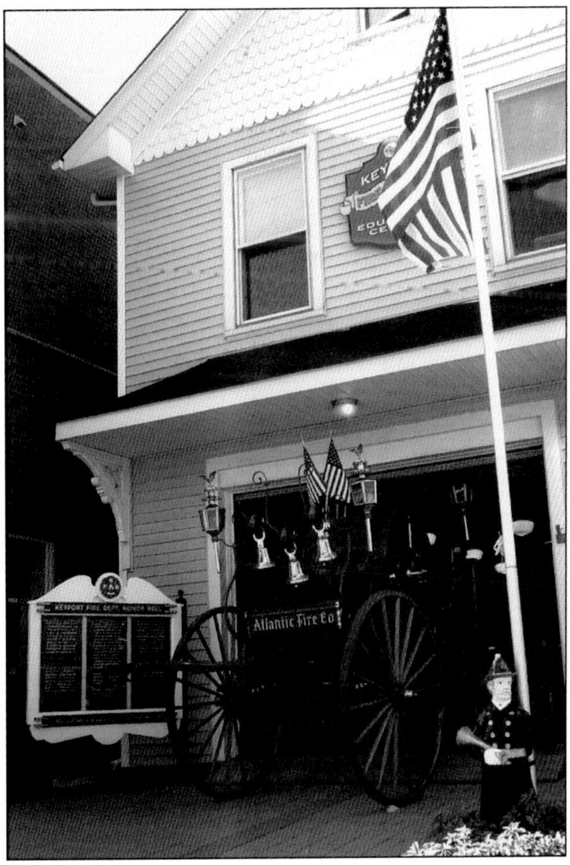

KEYPORT FIRE MUSEUM. In 2000, a group of history-minded firefighters and citizens banded together and officially incorporated as the Keyport Fire Museum and Education Center. With the assistance of then Keyport mayor John Merla, June Atkins, and Sen. Joseph Kyrillos, a state grant was obtained in 2002. At this time, the long-neglected former Raritan Hose Company firehouse was purchased. In 2003, after one year of countless volunteer man-hours and nearly $100,000 in donated materials, the museum opened as the first fire museum in Monmouth County.

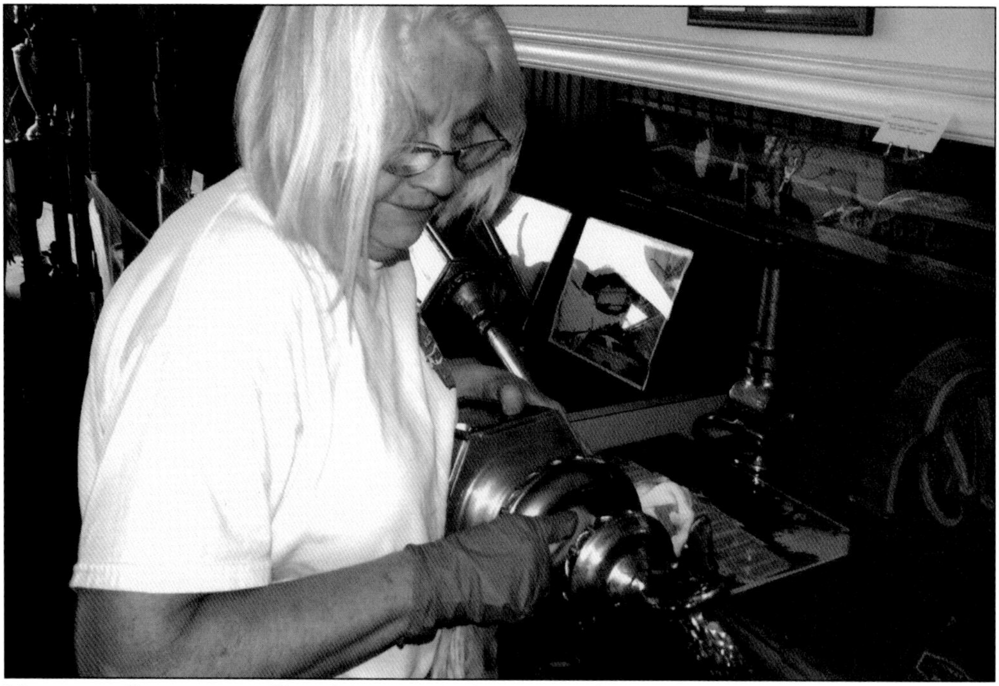

MARJORIE SECKINGER. Here Marjorie Seckinger, a charter member of the fire museum, is polishing a century-old signal lamp. Marjorie and her husband, Rowland, have dedicated thousands of hours to operate the museum and maintain the building since its opening in 2003. Her membership, skills, and sincerity have been an asset in reaching the museum's goals.

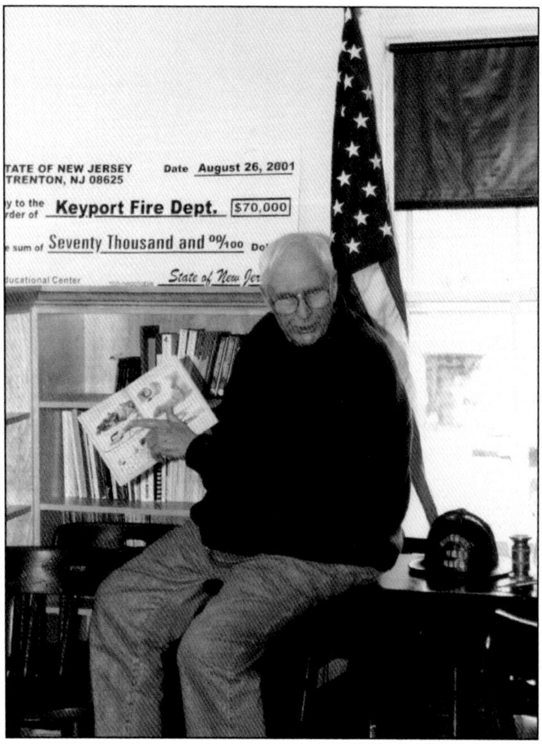

ROLAND SECKINGER. A native of Savannah, Georgia, Rowland and his wife retired to Keyport and soon became active in many volunteer projects. Rowland oversees many projects at the museum and has been on hand for his "southern hospitality" for schoolchildren and adults alike. He is seen here singing his favorite Jiminy Cricket fire safety story from a 1950s children's book.

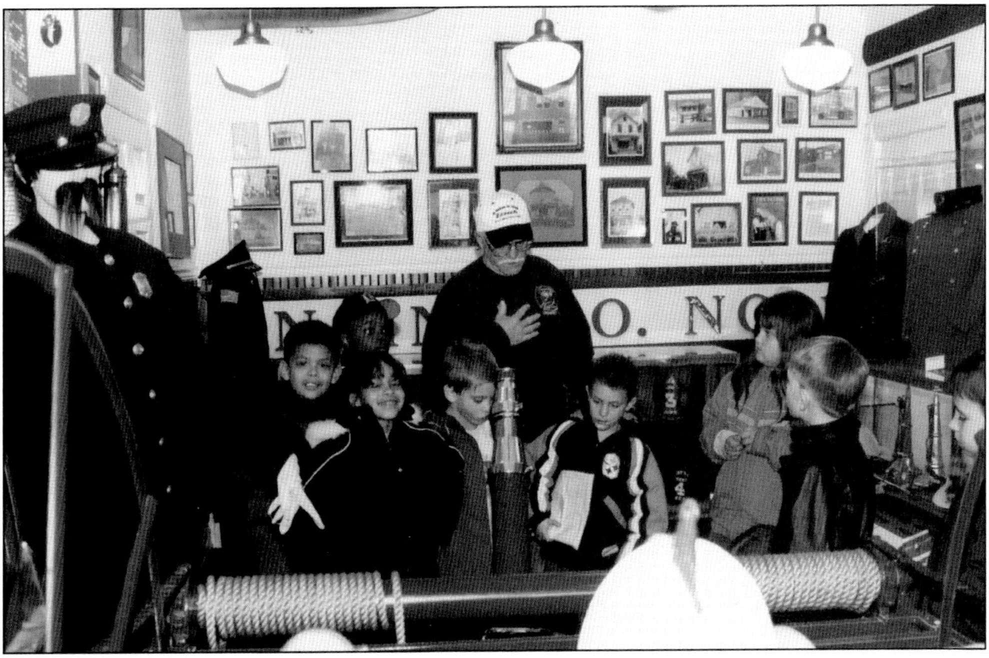

ROBERT POLING. Robert Poling has served as the secretary of the fire museum since its inception. He also oversaw the entire renovation project and spent nearly every day at the museum in the rough for over a year until the restoration was complete. He is seen here during a 2004 Keyport Central School class trip.

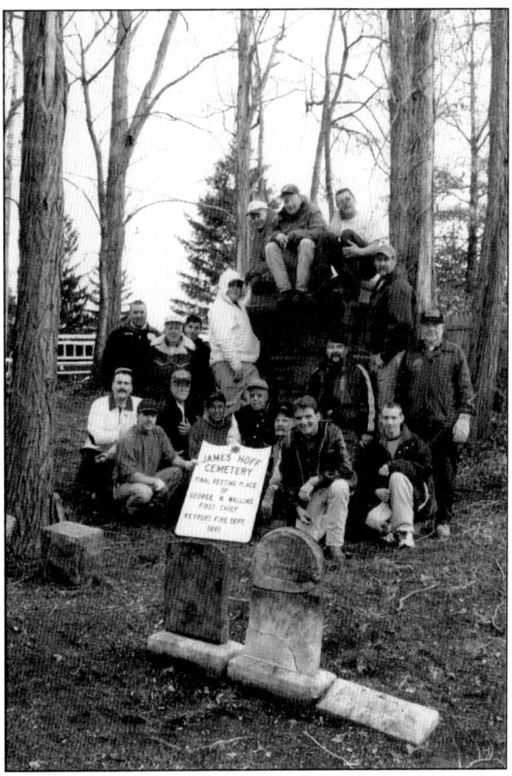

CEMETERY CLEANUP. One of the first service projects undertaken by the KFMA was this 2000 cemetery cleanup. The old Walling burial ground in Holmdel was completely cleaned and returned to a manageable condition after many years of neglect. This cemetery is the final resting place of George Washington Walling, the first chief of the KFD, who died of pneumonia in 1891.

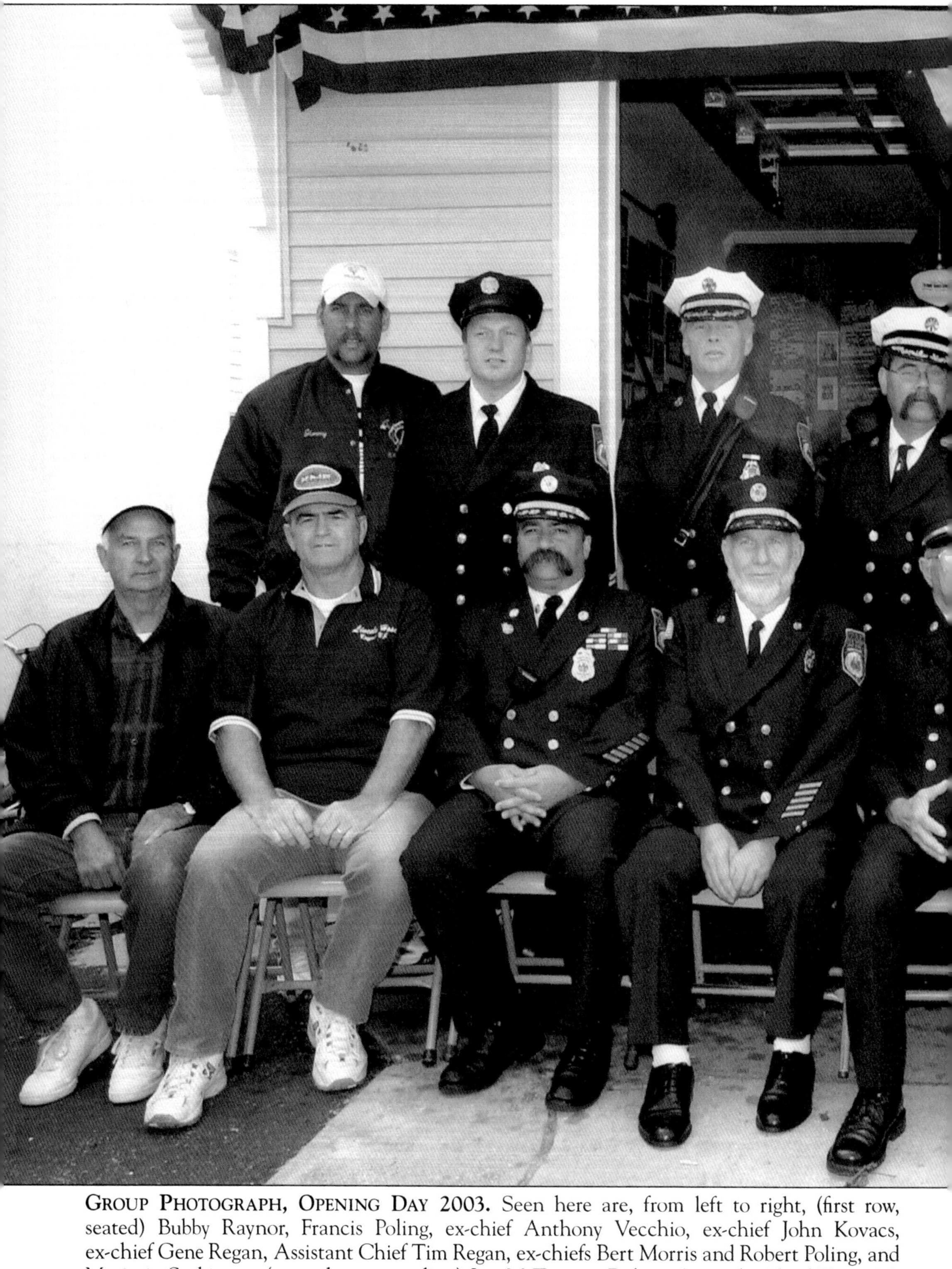

Group Photograph, Opening Day 2003. Seen here are, from left to right, (first row, seated) Bubby Raynor, Francis Poling, ex-chief Anthony Vecchio, ex-chief John Kovacs, ex-chief Gene Regan, Assistant Chief Tim Regan, ex-chiefs Bert Morris and Robert Poling, and Marjorie Seckinger; (second row, standing) Jim McTernan, Robert Aumack, Chief Harry M.

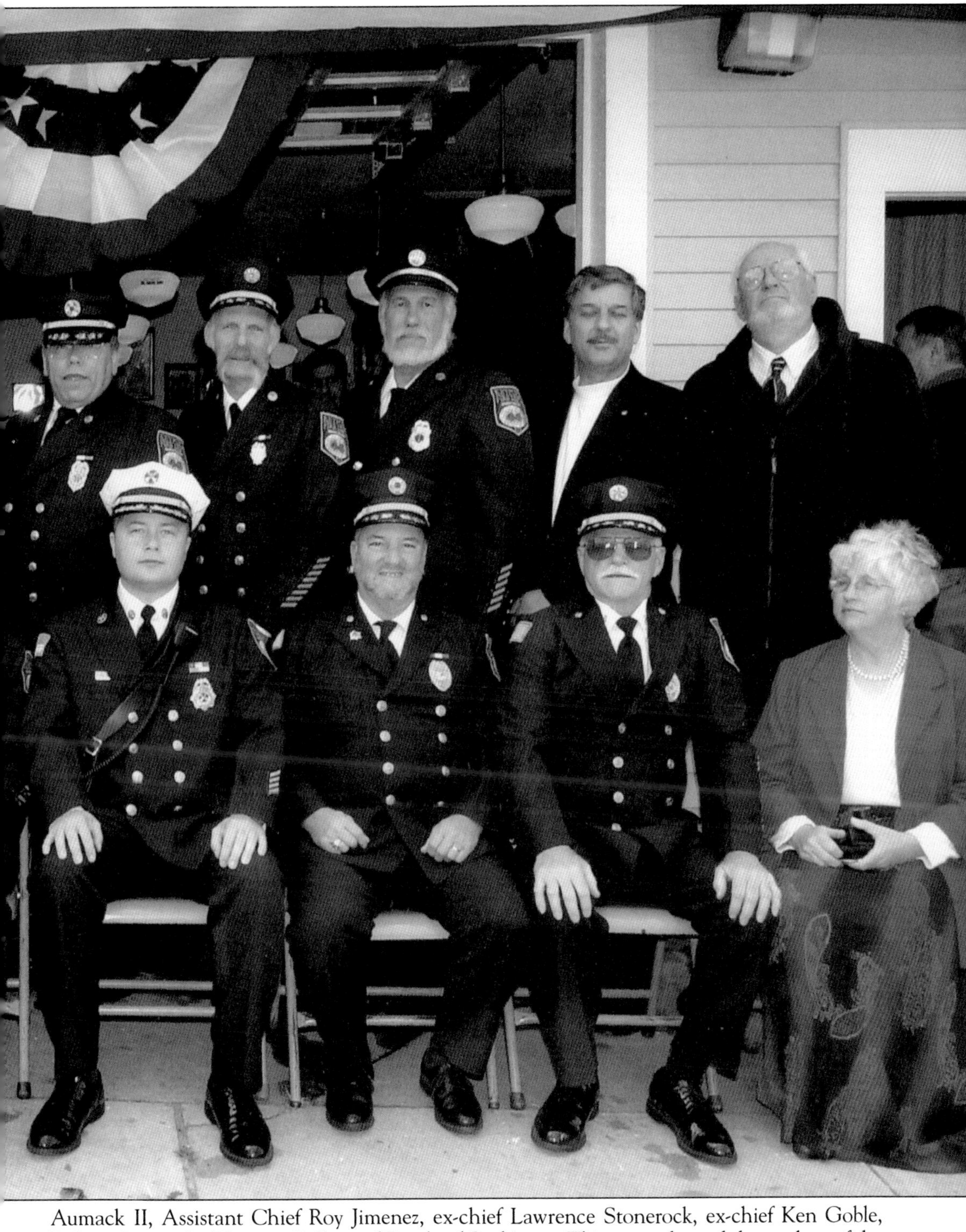

Aumack II, Assistant Chief Roy Jimenez, ex-chief Lawrence Stonerock, ex-chief Ken Goble, Robert Ciecierski, Thomas Gallo, and Rowland Seckinger. This group formed the nucleus of the staff, and several years later, the museum recorded annual visitation in excess of 1,700 persons.

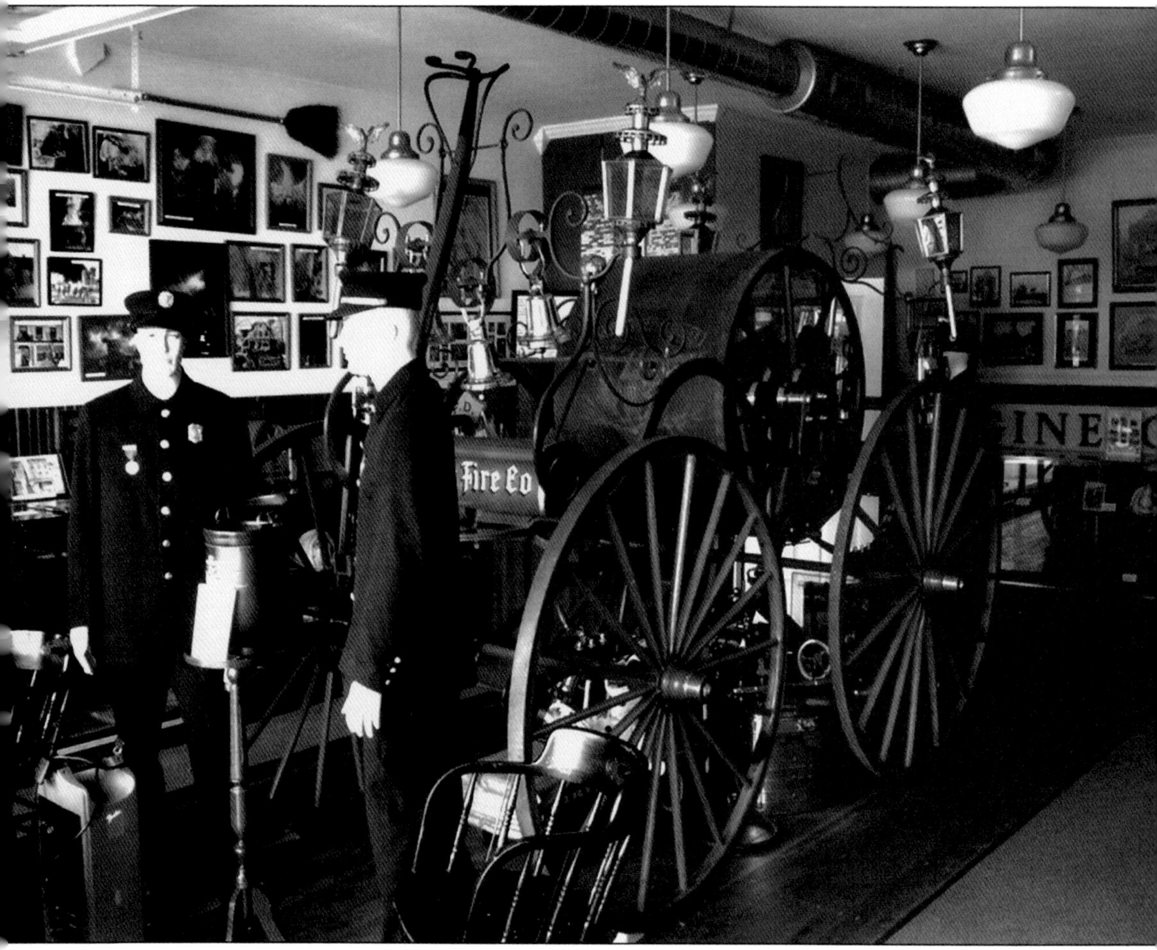

INTERIOR OF MUSEUM. Originally built in 1900, the old Raritan Hose Company firehouse was moved to its present location at 86 Broad Street in 1919. Today over 30,000 artifacts ranging from an 1880 hose carriage to a memorial built from the steel of the World Trade Center combine to make an interesting and educational trip for young and old.

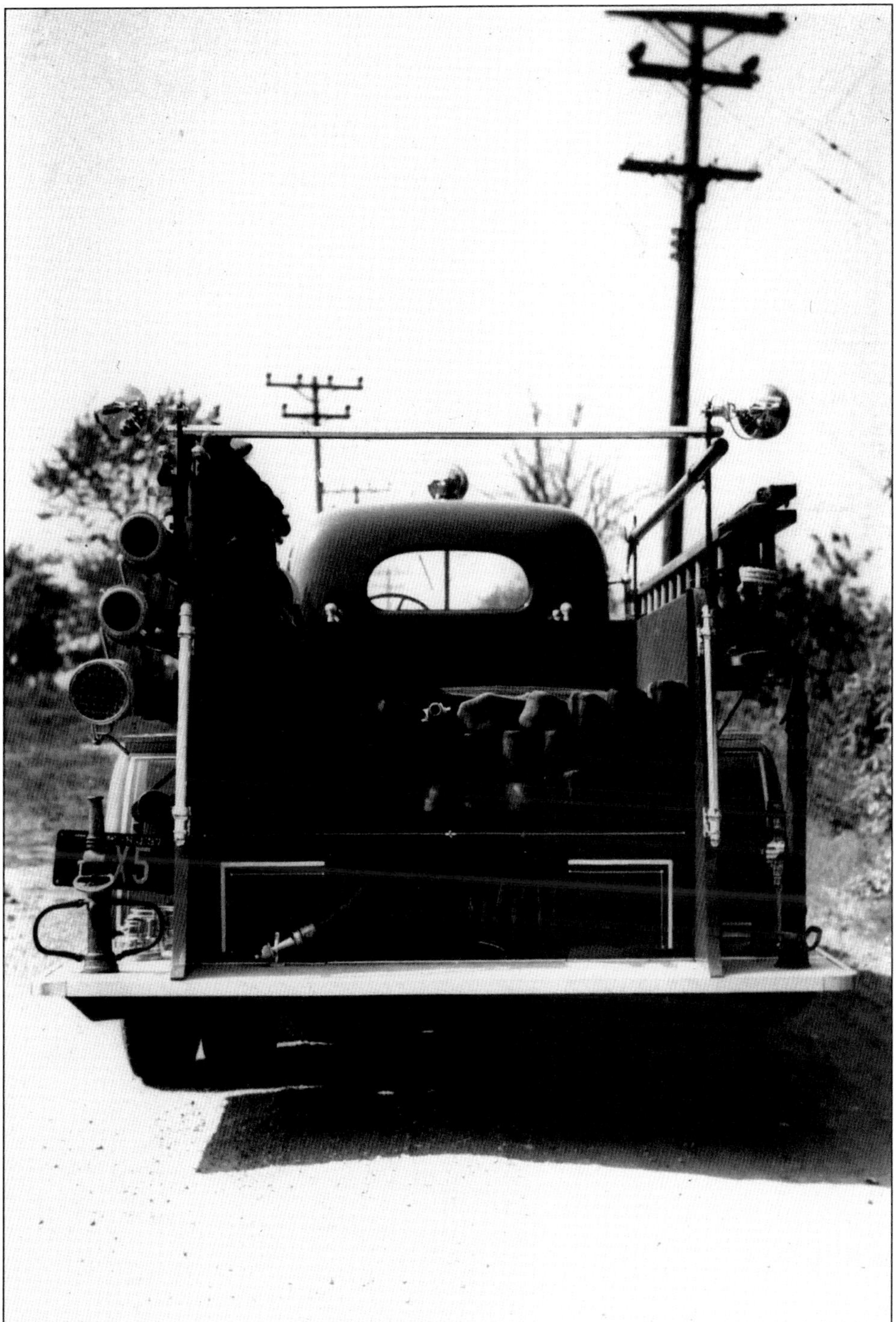

"THE FIRE IS OUT, ALL MEN AND EQUIPMENT ARE RETURNING." The firefighters will return home now, or clock back in at their day jobs. The department's volunteers are the unsung heroes of public service, willing to lay their lives on the line to protect life and property. The town is safe for now, until the siren sounds again.